AF746479

Dychweler y llyfr hwn erbyn y dyddiad isod
[illegible] gan y Llyfrgellydd

BIOCHEMICAL SOCIETY SYMPOSIA

No. 57

PROTEIN STRUCTURE, PREDICTION AND DESIGN

BIOCHEMICAL SOCIETY SYMPOSIUM No. 57
held at the University of Bath, April 1990

Protein Structure, Prediction and Design

ORGANIZED AND EDITED BY
J. KAY, G. G. LUNT AND D. J. OSGUTHORPE

1990
LONDON: THE BIOCHEMICAL SOCIETY

U.C.N.W. LIBRARY

SCIENCE
QH345.B522
no. 57

Published by Portland Press, 59 Portland Place, London W1N 3AJ on behalf of the Biochemical Society

Copyright © 1990 by the Biochemical Society

ISBN: 1 85578 002 X
ISSN: 0067 8694

British Library Cataloguing in Publication Data
Protein structure, prediction and design.
1. Proteins. Structure
I. Kay, John II. Lunt, George III. Osguthorpe, David
IV. Series
547.75044

ISBN 1-85578-002-X

All rights reserved
No part of this book may be reproduced in any form by photostat, microfilm or any other means, or stored in a retrieval system or transmitted in any form by any means, without written permission from the publishers. Copies of articles in this book may be made in the U.S.A. for personal or internal use, or for the personal or internal use of specific clients, on the condition, however, that the copier pay the stated per-copy fee through the Copyright Clearance Center Inc. (P.O. Box 765, Schenectady, New York 12301, U.S.A.) for copying beyond that permitted by Sections 107 or 108 of the U.S. Copyright Law. This consent does not extend to other kinds of copying, such as copying for general distribution, for advertising or promotional purposes, for creating new collective works, or for resale. The Copyright Clearance Center code for this book is 0067–8694/90/ $3.00 + 0.00.

Typeset by Santype International Limited, Salisbury, Wiltshire and Printed in Great Britain by Whitstable Litho Printers Ltd

215747

List of Contributors

J. Ball (*Proteus Molecular Design Ltd, 48 Stockport Road, Marple, Cheshire SK6 6AB, U.K.*)

V. Biou (*Unité d'Ingénierie des protéines, Biotechnologies, INRA 78352 Jouy-en-Josas, Cedex France*)

T. L. Blundell (*Laboratory of Molecular Biology and ICRF Unit of Structural Molecular Biology, Department of Crystallography, Birkbeck College, University of London, Malet Street, London WC1E 7HX, U.K.*)

L. Bülow (*Department of Pure and Applied Biochemistry, Chemical Centre, PO Box 124, S-211 00 Lund, Sweden*)

J. C. Cheetham (*Laboratory of Molecular Biophysics, University of Oxford, The Rex Richards Building, South Parks Road, Oxford OX1 3QU, U.K.*)

M. Claessens (*COMETT Technical Office, avenue de Cortenbergh, 71, B-1040 Brussels, Belgium*)

V. B. Cockcroft (*Molecular Graphics Unit and Biochemistry Department, University of Bath, Claverton Down, Bath BA2 7AY, U.K.*)

B. C. Cunningham (*Department of Biomolecular Chemistry, Genentech Inc., 460 Pt. San Bruno Blvd., South San Francisco, CA 94080, U.S.A.*)

F. Eisenmenger (*Laboratory of Molecular Biology and ICRF Unit of Structural Molecular Biology, Department of Crystallography, Birkbeck College, University of London, Malet Street, London WC1E 7HX, U.K.*)

J. B. C. Findlay (*Department of Biochemistry, University of Leeds, Leeds LS2 9JT, U.K.*)

F. V. Fishleigh (*Proteus Molecular Design Ltd, 48 Stockport Road, Marple, Cheshire SK6 6AB, U.K.*)

J. Garnier (*Unité d'Ingénierie des protéines, Biotechnologies, INRA 78352 Jouy-en-Josas, Cedex France*)

J. A. Gerlt (*Department of Chemistry and Biochemistry, University of Maryland, College Park, Maryland 20742, U.S.A.*)

J. F. Gibrat (*Unité d'Ingénierie des protéines, Biotechnologies, INRA 78352 Jouy-en-Josas, Cedex France*)

P. Greaney (*Proteus Molecular Design Ltd, 48 Stockport Road, Marple, Cheshire SK6 6AB, U.K.*)

D. S. Gregory (*Laboratory of Molecular Biophysics, University of Oxford, The Rex Richards Building, South Parks Road, Oxford OX1 3QU, U.K.*)

P. L. Howell (*Department of Chemistry, Massachusetts Institute of Technology, Cambridge, Massachusetts 02139, U.S.A.*)

M. S. Johnson (*Laboratory of Molecular Biology and ICRF Unit of Structural Molecular Biology, Department of Crystallography, Birkbeck College, University of London, Malet Street, London WC1E 7HX, U.K.*)

G. L. Kenyon (*Department of Pharmaceutical Chemistry, University of California, San Francisco, California 94143, U.S.A.*)

E. M. Kosower (*Biophysical Organic Chemistry Unit, School of Chemistry, Sackler Faculty of Exact Sciences, Tel-Aviv University, Tel-Aviv 69978, Israel*)

J. W. Kozarich (*Department of Chemistry and Biochemistry, University of Maryland, College Park, Maryland 20742, U.S.A.*)

I. Lasters (*Plant Genetic Systems CP160, P2, Universite Libre de Bruxelles, Ave. P. Heger, B-1050 Brussels, Belgium*)

J. M. Levin (*Unité d'Ingénierie des protéines, Biotechnologies, INRA 78352 Jouy-en-Josas, Cedex France*)

J. Li (*Proteus Molecular Design Ltd, 48 Stockport Road, Marple, Cheshire SK6 6AB, U.K.*)

G. G. Lunt (*Biochemistry Department, University of Bath, Claverton Down, Bath BA2 7AY, U.K.*)

A. Marsden (*Proteus Molecular Design Ltd, 48 Stockport Road, Marple, Cheshire SK6 6AB, U.K.*)

A. C. R. Martin (*National Institute for Medical Research, The Ridgeway, Mill Hill, London NW7 1AA, U.K.*)

D. C. Neidhart (*Department of Chemistry, Massachusetts Institute of Technology, Cambridge, Massachusetts 02139, U.S.A.*)

A. C. T. North (*Astbury Department of Biophysics, University of Leeds, Leeds LS2 9JT, U.K.*)

D. J. Osguthorpe (*Molecular Graphics Unit, University of Bath, Claverton Down, Bath BA2 7AY, U.K.*)

J. P. Overington (*Laboratory of Molecular Biology and ICRF Unit of Structural Molecular Biology, Department of Crystallography, Birkbeck College, University of London, Malet Street, London WC1E 7HX, U.K.*)

J. Pedersen (*Laboratory of Molecular Biophysics, University of Oxford, The Rex Richards Building, South Parks Road, Oxford OX1 3QU, U.K.*)

G. A. Petsko (*Department of Chemistry, Massachusetts Institute of Technology, Cambridge, Massachusetts 02139, U.S.A.*)

F. Pio (*University of California, Department of Medicine, Division of Endocrinology and Metabolism, San Diego, La Jolla, CA 92093, U.S.A.*)

E. Platt (*Proteus Molecular Design Ltd, 48 Stockport Road, Marple, Cheshire SK6 6AB, U.K.*)

J. L. Pool (*Proteus Molecular Design Ltd, 48 Stockport Road, Marple, Cheshire SK6 6AB, U.K.*)

V. M. Powers (*Department of Pharmaceutical Chemistry, University of California, San Francisco, California 94143, U.S.A.*)

A. R. Rees (*Department of Biochemistry, University of Bath, Claverton Down, Bath BA2 7AY, U.K.*)

B. Robson (*Proteus Molecular Design Ltd, 48 Stockport Road, Marple, Cheshire SK6 6AB, U.K.*)

C. Sander (*E.M.B.L., Postfach 10.2209, Meyerhofstrasse 1, D-6900 Heidelberg, F.R.G.*)

J. Saunders (*Merck, Sharpe & Dohme Research Laboratories, Neuroscience Research Centre, Terlings Park, Eastwick Road, Harlow, Essex CM20 2QR, U.K.*

D. Staunton (*Laboratory of Molecular Biophysics, University of Oxford, The Rex Richards Building, South Parks Road, Oxford OX1 3QU, U.K.*)

P. Thomas (*Laboratory of Molecular Biology and ICRF Unit of Structural Molecular Biology, Department of Crystallography, Birkbeck College, University of London, Malet Street, London WC1E 7HX, U.K.*)

C. M. Topham (*Laboratory of Molecular Biology and ICRF Unit of Structural Molecular Biology, Department of Crystallography, Birkbeck College, University of London, Malet Street, London WC1E 7HX, U.K.*)

J. A. Wells (*Department of Biomolecular Chemistry, Genentech, Inc., 460 Pt. San Bruno Blvd., South San Francisco, CA 94080, U.S.A.*)

S. J. Wodak (*Unite de Conformation de Macromolecules Biologiques, CP160, P2, Universite Libre de Bruxelles, Ave. P. Heger, B-1050 Brussels, Belgium*)

Preface

The 634th meeting of the Biochemical Society was held in the University of Bath, from 2–6 April. It was a happy coincidence that the meeting coincided with the arrival in Bath of the new Head of the Biochemistry Department, Professor A. R. Rees and the establishment of a new protein crystallography facility. This had not been part of the original planning of the meeting! An important feature of Biochemistry at Bath is the close interaction with the Department of Chemistry and this is particularly evident in the area of molecular graphics. Thus we took the central theme for this annual symposium from one of the major questions facing both our own colleagues in Bath and many others elsewhere, namely: How can we interpret and exploit the ever-increasing body of knowledge about the primary structure of proteins in terms of tertiary structure and, ultimately, function?

This year's annual Symposium brought together some of the world's leading experts in protein structure and its manipulation to present the current state-of-the-art in this field. Although the fundamental problem of protein folding is not yet solved, that is, given an arbitrary amino acid sequence no one can yet predict its structure with a high degree of reliability, the Symposium presented current attempts to determine ways of predicting protein structure and to manipulate protein structure. The Symposium addressed three main areas, protein structure, prediction and design.

The first section concentrated on what has been learned from known protein structures and the use of these data in protein structure prediction algorithms, concentrating on secondary structure prediction and homology modelling. Homology modelling is the most widespread technique for generating structures for proteins whose sequence is known. It is based on using a protein of known structure as a template for the folding of a protein of similar sequence. Extensions to these algorithms were described which introduce additional information about the sequence, e.g. conservation of residue type, flexible or rigid regions and others. These extensions improve the ability of homology modelling to attack sequences which have lower similarity to the sequence of known structure.

The recent determination of the single membrane protein structure known to date, the photosynthetic reaction centre, highlights our poor knowledge of the structure of proteins that traverse or sit in the cell membrane. Yet many of the most important proteins from the viewpoint of understanding cellular signalling systems are transmembrane proteins. The prediction of the structure of these proteins is at a very early stage but, due to the difficulty of obtaining their structures by classical techniques, structural modelling of these proteins can aid in the analysis of experimental data. This area was covered by two talks, one on acetylcholine receptor modelling and another on second messenger-linked receptor structures.

Although given a sequence we cannot easily find its structure, the inverse process may be less difficult, that is to say, given a structure can we design a sequence that will adopt this fold? Creating protein sequences and producing them for an experimental structure determination is being attempted by a few laboratories and has been achieved.

The final section of the symposium considered attempts to modify the activity of enzymes or to create new activities. This considered techniques for assembling enzyme complexes from individual enzymes by fusing individual proteins into a single polypeptide chain. The final presentations considered the production of enzyme activities via catalytic antibodies. This strategy circumvents our current inability to design such activities *ab initio* and instead utilizes the ability of the immune system to create proteins which will bind specific ligands. By chemically creating transition state analogues for the reaction of interest the immune system will create proteins which bind these analogues. This has been achieved and we saw how a number of enzymes have been created for novel reactions.

We hope this volume will help to recapture some of the excitement of the symposium and that it will serve to show those outside the field that the prediction and design of protein structure and function is now an attainable goal.

JOHN KAY
GEORGE LUNT
DAVID OSGUTHORPE

Cardiff and Bath

Contents

Abbreviations

ACh	Acetylcholine
AChR	Acetylcholine receptor
BLG	β-Lactoglobulin
B2C	Flavocytochrome b_2
CDR	Complementarity-determining region
cGMP-NaC	Cyclic GMP-gated sodium channel
Con A	Concanavalin A
FABP	Fatty acid binding protein
FPS	Functional programming system
GABA	γ-Aminobutyric acid
GLO	Glycolate oxidase
6-HDNO	6-Hydroxy-D-nicotine oxidase
hGH	Human growth hormone
IGPS	Indoglycerol phosphate synthase
KGA	Aldolase
LGIC	Ligand-gated ion-channel
Lmrop	Left-handed monomer of rop
MIR	Main immunogenic region
MLE	Muconate lactonizing enzyme
MR	Mandelate racemase
MR	Muscarinic receptor
NaC	Sodium channel
nAChR	Nicotinic acetylcholine receptor
NMS	*N*-Methylscopolamine
OBP	Odorant-binding protein
OXO-M	Oxotremorine-M
PPase	Pyrophophatase structure
PRAI	Phosphoribosylanthranilate isomerase
PtdIns	Phosphatidylinositol
PYK	Pyruvate kinase
RBP	Retinol-binding protein
Rh	Rhodopsin
rop	Repressor of primer
SGR	Single group rotation
TAA	Taka-amylase
TEADH	Triethylamine dehydrogenase
THIP	4,5,6,7-Tetrahydroisoxazolo[5,4*c*]pyridin-3-ol
TIM	Triose phosphate isomerase
Wtrop	Wild-type rop

Biochem. Soc. Symp. **57**, 1–9
Printed in Great Britain

An Assessment of COMPOSER: a Rule-Based Approach to Modelling Protein Structure

C. M. TOPHAM, P. THOMAS, J. P. OVERINGTON, M. S. JOHNSON, F. EISENMENGER and T. L. BLUNDELL

Laboratory of Molecular Biology and I.C.R.F. Unit of Structural Molecular Biology, Department of Crystallography, Birkbeck College, University of London, Malet Street, London WC1E 7HX, U.K.

Summary

We describe COMPOSER, a computer program for modelling proteins which uses three-dimensional structures defined by X-ray analysis together with rules defined by their analysis and comparison. We discuss the modelling of phospholipases and serine proteinases of known structure to assess the errors associated with such models.

Introduction

Recent structure determinations suggest that the majority of new protein structures are comprised of motifs or domains that have been previously identified in other, often functionally different, proteins. If we can identify the protein fold that is assumed by an amino acid sequence and if we can define rules for a family of proteins that can be extrapolated to the new sequence, then we have a feasible route to modelling a protein from its sequence.

The classical approach to this problem was developed for modelling one protein on the three-dimensional structure of an homologue. The first application was the construction of a model for α-lactalbumin on the basis of lysozyme (Browne *et al.*, 1969). Other applications included the construction of models for relaxins and insulin-like growth factors (see Blundell & Humbel, 1980, for a review) and various serine proteinases (Greer, 1981). Some rules for modelling protein structure (Greer, 1981; Sibanda & Thornton, 1985; Sibanda *et al.*, 1989) have been suggested and the advent of computational techniques, particularly of the computer graphics program FRODO (Jones & Thirup, 1986), has made modelling more straightforward.

In this paper we shall review an approach to protein modelling that is completely automated and in which all decisions are rule-based (Blundell *et al.*, 1987; Sali *et al.*, 1990). We begin by comparing or aligning protein sequences and three-dimensional structures determined by experiment and organized in databases. We use these comparisons to derive rules about families of protein structures that adopt a common fold. The rules are then used to define a template for sequences for each family fold. If a sequence of protein of unknown structure can be matched with a template, we can then extrapolate features of the fold to the sequence and so construct a model of the protein. Thus, we exploit not only the details of protein structures, but also the rules developed from their analysis and comparison; these together form the knowledge base.

In this paper we discuss a simple approach to knowledge-based modelling of proteins which is encoded in COMPOSER (Sutcliffe *et al.*, 1987*a,b*; Blundell *et al.*, 1988). We discuss models that have been constructed for proteins the structures of which have also been defined experimentally, and describe a useful estimate of error in the models.

COMPOSER

COMPOSER is a suite of computer programs for the knowledge-based modelling of the three-dimensional structure of a protein from the known tertiary structures of homologous proteins. This method depends on the assembly of three sets of fragments of protein structures (Sutcliffe *et al.*, 1987*a,b*; Blundell *et al.*, 1988). These sets of fragments are: (1) helices and strands of the framework (Sutcliffe *et al.*, 1987*a*); (2) regions outside the framework selected from a database of loop sub-structures using a distance filter in a similar way to Jones & Thirup (1986) and (3) side-chains selected using rules derived from the analysis of homologous structures (Sutcliffe *et al.*, 1987*b*).

The first stage is the simultaneous rigid-body superposition of the known structures using an iteratively reweighted least-squares procedure to define the topologically equivalent residues (Sutcliffe *et al.*, 1987*a*). This common core or framework comprises a series of helices and/or strands with occasional connecting turns that are conserved in the family.

Once protein structures have been aligned or equivalenced, clustering methods using a matrix of similarity scores between all pairs of proteins can be used to construct a tree that describes relationships between them. While there has been much discussion of methods for construction of evolutionary trees from sequences, the clustering of protein three-dimensional structures has been less studied. Eventoff & Rossmann (1975) constructed dendrograms based on structural features alone to describe distant phylogenetic relationships among the mono- and di-nucleotide-binding proteins. We have shown that a useful measure of structural difference can be obtained from differences between superposed structures (Johnson *et al.*, 1990*a*). We have extended this approach by considering additional structural and sequence features in the comparisons (Sali & Blundell, 1990; Johnson *et al.*, 1990*b*), so that proteins with little sequence similarity can be compared and classified. Such structural trees can be used in conjunction with the sequence-based trees to provide automatic selection of the best structures and fragments for modelling from sequence.

The framework for modelling is calculated as the weighted mean for each equivalent C_α position. The sequence of the unknown is then aligned to a template that represents the sequences of the known structures in the framework region. Fragments of the proteins of known structure that have the closest similarity to the unknown are least-squares fitted to this structurally conserved core region to give the model real protein geometry. Finally, the side-chains are added using a set of environment-dependent rules (Sutcliffe *et al.*, 1987*b*).

The next stage is to model the variable regions which connect the framework elements of secondary structure. Protein fragments which are of the same length as the loop to be modelled and which are compatible with the geometry of a three-residue overlap with the framework are first identified by searching a local distance matrix database of all structures including members of the same family. Several types of filter may then be successively applied in the selection of the most appropriate loop fragment to be annealed to the model core. We use a novel ring closure algorithm in which the main-chain torsion angles of the loop are tweaked causing minimal disturbance to the structurally conserved core (F. Eisenmenger, unpublished work). Geometric filters dependent on initial differences in *N*- and *C*-terminal overlaps or changes in the torsion angles and r.m.s. (root mean square) differences in the C_α positions in the loop during ring closure may be used.

We are currently testing a selection procedure which generates a template for each fragment selected on geometric criteria. It is based on prediction of the amino acid substitutions consistent with the conformation of each fragment. Each amino acid can belong to one of 64 different physical environ-

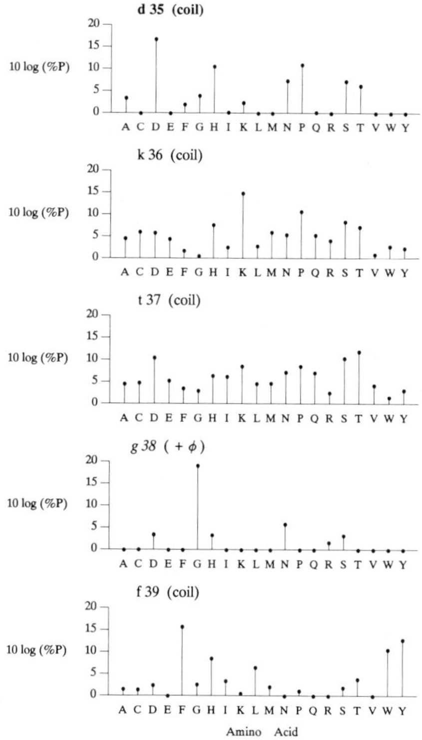

Fig. 1. *Substitution patterns predicted for a β-hairpin of class 3 : 5 from chymotrypsin* Sequence (35–39) : Asp-Lys-Thr-Gly-Phe.

ments. For each of these the substitution pattern has been estimated by comparing residues at topologically equivalent positions in families of homologous proteins (Overington *et al.*, 1990). Fig. 1 indicates the predicted substitution pattern obtained using a subset of the mutation matrices described by Overington *et al.* (1990), created by scoring only those pairwise substitutions in which the secondary structure class is observed, for a class 3 : 5 β-hairpin bend (Sibanda *et al.*, 1989) from chymotrypsin. As can be seen, there is a specific requirement for the residue at the fourth position to be a glycine, whereas there are relatively fewer constraints on the identities of the remaining residues.

In some cases there will be steric clashes between the fragments chosen. Very often these can be relaxed using energy minimization procedures, but often the next most probable fragment must be selected instead of the first choice.

Modelling Known Structures

We have used COMPOSER to model many different protein structures (see, for example, Sutcliffe *et al.*, 1987*a*, 1987*b*, Overington *et al.*, 1989, and Topham *et al.*, 1990). However, to assess the errors that might be expected through the use of such a procedure, we have used COMPOSER to build models of known three-dimensional protein structures.

Phospholipases

There are three available X-ray crystallographic structures of members of the phospholipase A_2 family: the porcine (1p2p; Dijkstra *et al.*, 1983) and bovine (1bp2; Dijkstra *et al.*, 1981) pancreatic monomeric enzymes, and the diamondback rattlesnake (1pp2; Brunie *et al.*, 1985) dimeric enzyme. We have compared these three-dimensional structures using COMPARER (Sali & Blundell, 1990); this program equivalences similar structural features in the proteins and produces an alignment of the sequences based on this comparison (Fig. 2). The nature of the equivalenced structural features such as inaccessible side-chain or side-chain hydrogen bonds are indicated schematically.

We have modelled the porcine enzyme on the bovine and rattlesnake enzyme structures. These have respectively 87% and 47% sequence identities with the porcine enzyme in a framework region of 88 amino acid residues. The resulting model is shown in Fig. 3 compared with the experimentally defined structure and has an r.m.s.d. of 1.46 Å for all the main-chain C_αs, but this falls dramatically to 0.63 Å if residues 56–66 in the common numbering system (Renetseder *et al.*, 1985) shown in Fig. 2 are excluded. Comparable r.m.s.d. values of 1.44 Å and 0.62 Å were obtained for the model of the bovine enzyme built from the other pair of known structures. Given the high level of sequence identity between the bovine and porcine phospholipases A_2 in the loop region 56–68, it is surprising that the two very different conformations are seen in the crystal structures in this region and, indeed, recent evidence from ^{1}H n.m.r.

```
                      10                   20                   30                   40                 50
p1bp2   a l w Q f n g M I k̃ C k̃ i p s S̃ ẽ P l l d̃ f n ñ Ỹ G C y C g l g G s g t p v d̃ d L D r̃ C C q̃ t̃ h̃ d n
p1p2p   a l w Q f r s̃ M I k c a i p g s h P l m d f n ñ Ỹ G C y C g l g G s̃ g t p v d̃ e L D r̃ C C ẽ t̃ H̃ d n
p1pp2   s l v q f e t L I m k i A - g r s G l l w Y s a Ỹ G C Y C g w G g h G l P q̃ d̃ a T D r̃ C c f v h̃ d c
          h h h h h h h h h h h                          +            ++  +          h h h h h h h h h h h
        ===========     ===============     ===============

                      60                   70                   80                   90                 100
p1bp2   c ỹ k q̃ A k k̃ l d s C k v l v d̃ ñ P y t̃ Ñ n Ỹ s y s c s n n ẽ I t c s s e Ñ n a C ẽ a f I C n C D̃ r
p1p2p   c Ỹ r d A k n l d s C k F l v d n P ỹ t e s Ỹ s̃ ỹ s c s n t e I t c ñ s k̃ Ñ ñ a c ẽ a f I C n C D̃ r
p1pp2   c ỹ g k A t - - d - c - - - - - ñ p k t̃ v s Ỹ t y s e e n g e I i C g G d - d p c G t q I C e C D̃ k
        h h                                               b b b  ++  b b b        h h h h h h h h h h h
        =====                        =======       ===       ===========

                      110                  120      126           133
p1bp2   n A A i c F s k v - p - y ñ k e h̃ k ñ l d k - k n C
p1p2p   n A A i c F s k A - p - ỹ ñ k e h̃ k̃ ñ l d t k̃ k y C
p1pp2   a A A i c F r d n i p s Ỹ d̃ n k̃ y w l f p p - k d C r ẽ e p ẽ p c
        h h h h h h h            h h h  +
        =========      =========
```

Fig. 2. *An alignment of the sequences of the phospholipase A_2 family obtained by comparing their three-dimensional structures using the method of Sali & Blundell (1990)*

The consensus secondary structure (h: helix; b: sheet) and structurally equivalent residues are indicated =. Upper case, solvent inaccessible; lower case, solvent accessible; underlined, H-bond from side-chain to main-chain CO; bold, H-bond from side-chain to main-chain NH; tilde (˜), H-bond from side-chain to other side-chain; italic, positive ϕ main-chain torsion angle.

studies (Fisher *et al.*, 1989) indicates that this loop in fact adopts the same conformation in the two enzymes in solution. The problem may arise from crystal packing effects or, more likely in this case, errors in the X-ray analysis. Clearly the lower r.m.s.d. value of approx. 0.6 Å provides a better indication of the errors in the models of the bovine and porcine enzymes. Model building of the rattlesnake enzyme from the two known mammalian structures yielded a structure with an r.m.s.d. of 1.68 Å for the main-chain C_αs. The peptide fragments used to build the model core, comprising 79 residues, and the origins of the structurally variable regions selected from the database, are indicated in Table 1.

Serine proteinases

We have also constructed a model of bovine trypsin from the three-dimensional structures of other serine proteinases (Overington *et al.*, 1989). In this case there are more homologous structures available for use. There are eight useful serine proteinase structures in the latest release of Brookhaven Data Bank (Bernstein *et al.*, 1977). As expected, the r.m.s.d. differences between the model and the experimentally defined structure decreases as more structures are added in the construction of the framework. In this case the r.m.s. error in the model is 0.60 Å for the 152 core C_α and 0.63 Å for all the core main-chain atoms.

Owing to the larger number of available homologous structures, all but one of the twelve loop regions may be built directly from appropriate fragments in the homologous family. With all loop regions fitted to the core, the r.m.s.d. of

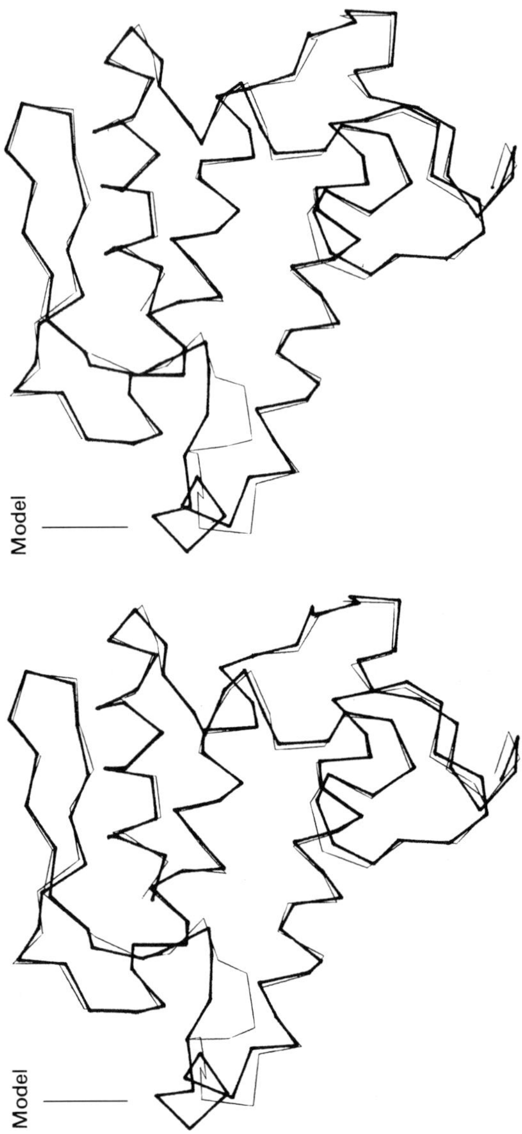

Fig. 3. *A model of porcine phospholipase produced using COMPOSER (bolder line) from homologous phospholipase structures and compared with the structure defined by X-ray analysis*

Table 1. *Protein fragments used in the modelling of rattlesnake phospholipase A2*

A consecutive numbering scheme is used here for the rattlesnake phospholipase A_2 structure. The other structures are numbered according to the coordinate files deposited in the Brookhaven Data Base (Bernstein *et al.*, 1977) and identified by four-letter codes shown in parentheses.

Structurally conserved region	Loop region	Loop sequence
Residues (1–12)	Residues (13–15)	A G R
Bovine phospholipase A_2 (1BP2) residues (1–12)	Scorpion neurotoxin (1SN3) residues (10–12)	D G C
Residues (16–28)	Residues (29–33)	G W G G H
Bovine phospholipase A_2 (1BP2) residues (17–29)	Bovine phospholipase A_2 (1BP2) residues (30–34)	G L G G S
Residues (34–54)	Residues (55–61)	T D C N P K T
Bovine phospholipase A_2 (1BP2) residues (35–55)	Tyrosyl tRNA synthetase (2TS1) residues (71–77)	G A T G L I G
Residues (62–66)	Residues (67–72)	S E E N G E
Porcine phospholipase A_2 (1P2P) residues (71–75)	Porcine phospholipase A_2 (1P2P) residues (76–81)	S C S N T E
Residues (73–75)	Residues (76–79)	G G D D
Bovine phospholipase A_2 (1BP2) residues (82–84)	γ-Chymotrypsin (2GCH) residues (45–48)	S L I N
Residues (80–97)	Residues (98–103)	D N I P S Y
Bovine phospholipase A_2 (1BP2) residues (90–107)	Photosynthetic reaction centre (1PRC M) residues (86–91)	R Q F F W L
Residues (104–110)	Residues (111–115)	P P K D C
Bovine phospholipase A_2 (1BP2) residues (112–118)	Bovine phospholipase A_2 (1BP2) residues (119–123)	D K K N C

the final model compared with the experimentally defined structure is 0.90 Å for all C_α atoms, although some regions have local errors of up to 4.5 Å. This worst part of the model is for the loop region which was modelled from a peptide fragment outside of the family.

Estimating the Errors in Models

It would be useful to have some indication of the likely errors in the final model as a function of residue position in the structure. A good indicator of this appears to be the contact number proposed by Oobatake & Ooi (1977). This is simply a count of the number of residues within a sphere of the atom considered. In this case the radius of the sphere was taken as 14 Å. As can be seen from Fig. 4 for the model building of the rattlesnake phospholipase A_2, this measure is inversely related to the r.m.s. error in the model. Similar results were obtained for the final trypsin model (correlation coefficient of -0.72); so, where there are few neighbouring residues the likely error will be higher than where there are many surrounding residues. This can be used to estimate the error, and the values might be used to estimate a disorder or temperature factor comparable to that defined experimentally in X-ray analysis.

C.M.T. is supported by a grant from the European Commission, M.S.J. is funded by the American Cancer Society, P.T. by Tripos Associates, F.E. by the Imperial Cancer

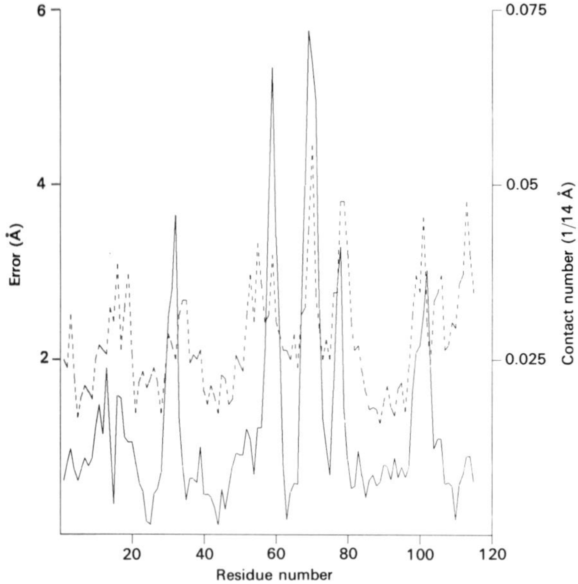

Fig. 4. *Relation between contact number and error*

Solid curve: errors in the position of the C_α atoms in the model of the rattlesnake phospholipase A_2. Dashed curve: the reciprocal 14 Å contact number for each C_α in the final model. Note the high correlation of this value and the errors in the model.

Research Fund and J.P.O. by Pfizer and S.E.R.C. We are grateful to the Imperial Cancer Research Fund and the S.E.R.C. for general financial support. We thank Dr Lynn Sibanda and Dr Janet Thornton for supplying data on their classifications of hairpin structures in proteins. We are grateful to our colleagues at Birkbeck for many helpful discussion.

References

Bernstein, F. C., Koetzle, T. F., Williams, G. J. B., Meyer, E. F., Brice, M. D., Rodgers, J. R., Kennard, O., Shimanouichi, T. & Tasumi, M. (1977) *J. Mol. Biol.* **112**, 535–542

Blundell, T. L. & Humbel, R. E. (1980) *Nature* (*London*) **287**, 781–787

Blundell, T., Carney, D., Gardener, S., Hayes, F., Howlin, B., Hubbard, T., Overington, J., Singh, D. A., Sibanda, B. L. & Sutcliffe, M. (1988) *Eur. J. Biochem.* **172**, 513–520

Blundell, T. L., Sibanda, B. L., Sternberg, M. J. E. & Thornton, J. M. (1987) *Nature* (*London*) **326**, 347–352

Browne, W. J., North, A. C. T., Phillips, D. C., Brew, K., Vanaman, T. C. & Hill, R. L. (1969) *J. Mol. Biol.*, **42**, 65–86

Brunie, S., Bolin, J., Gerwirth, D. & Sigler P. B. (1985) *J. Biol. Chem.* **260**, 9742–9749

Dijkstra, B. W., Kalk, J. H., Hol, W. G. J. & Drenth, J. (1981) *J. Mol. Biol.* **147**, 97–123

Dijkstra, B. W., Renelseder, R., Kalk, J. H., Hol, W. G. J. & Drenth, J. (1983) *J. Mol. Biol.* **168**, 163–179.

Eventoff, W. & Rossmann, M. G. (1975) *CRC Crit. Rev. Biochem.* **3**, 111–140

Fisher, J., Primrose, W. U., Roberts, G. C. K., Dekker, N., Boelens, R., Kaptein, R. & Slotboom, A. J. (1989) *Biochemistry* **28**, 5939–5946

Greer, J. (1981) *J. Mol. Biol.* **153**, 1027–1042

Johnson, M. S., Sutcliffe, M. J. & Blundell, T. L. (1990*a*) *J. Mol. Evol.* **30**, 43–49

Johnson, M. S., Sali, A. & Blundell, T. L. (1990*b*) *Methods Enzymol.* **183**, 670–690

Jones, T. H. & Thirup, S. (1986) *EMBO J.* **5**, 819–822

Oobatake, M. & Ooi, T. (1977) *J. Theor. Biol.* **67**, 567–584

Overington, J. P., Sutcliffe, M. J., Watson, F., James, K., Campbell, S. & Blundell, T. L. (1989) in *Proc. Int. Biotechnol. Symp.* (Durand, G., Bobichon, L. & Florent, J., eds.), pp. 279–304, Société Française de Microbiologie

Overington, J., Johnson, M., Sali, A. & Blundell, T. L. (1990) *Proc. R. Soc. London B*, **241**, 132–145

Renetseder, R., Brunie, S., Dijkstra, B. W., Drenth, J. & Sigler, P. B. (1985) *J. Biol. Chem.* **260**, 11627–11634

Sali, A. & Blundell, T. L. (1990) *J. Mol. Biol.* **212**, 403–428

Sali, A., Overington, J. P., Johnson, M. S. & Blundell, T. L. (1990) *Trends Biochem. Sci.* **15**, 235–240

Sibanda, B. L. & Thornton, J. M. (1985) *Nature* (*London*) **316**, 170–174

Sibanda, B. L., Blundell, T. L. & Thornton, J. M. (1989) *J. Mol. Biol.* **206**, 759–777

Sutcliffe, M. J., Haneef, I., Carney, D. & Blundell, T. L. (1987*a*) *Protein Eng.* **1**, 377–384

Sutcliffe, M. J., Hayes, F. R. F. & Blundell, T. L. (1987*b*) *Protein Eng.* **1**, 385–392

Topham, C. M., Overington, J., Kowlessur, D., Thomas, M., Thomas, E. W. & Brocklehurst, K. (1990) *Biochem. Soc. Trans.* **18**, 579–580

Biochem. Soc. Symp. **57**, 11–24
Printed in Great Britain

Secondary Structure Prediction and Protein Design

J. GARNIER, J. M. LEVIN, J. F. GIBRAT and V. BIOU

Unité d'Ingénierie des protéines, Biotechnologies, INRA 78352 Jouy-en-Josas, Cedex France

Synopsis

For non-homologous proteins, and after cross-validation, the methods reviewed in this article exhibit a probability index (percentage of correctly predicted residues per predicted residues) of 59–65.5% according to the methods employed with a standard deviation of 7% for three conformational states: α-helix, β-strand and coil. These present limitations in the accuracy of prediction are related both to the limited number of known structures and to the effect of long-range interactions.

The methods based on sequence similarity can improve the accuracy of prediction by expressing the homology of the protein to be predicted explicitly with proteins in the database. Under these circumstances, the probability index can reach 87% with a standard deviation of 6.6%. This property was used for modelling homologous proteins by assisting in amino acid sequence alignments. Examples will be given for the alignment of serine proteinases.

Introduction

Proteins are a remarkable example of a self-structuring system, even if one allows for the existence of chaperonin proteins which facilitate their folding [1]. The ability to decipher the code which relates the amino acid sequence of a protein to its three-dimensional structure remains a major goal in molecular biology, since proteins are at the root of the structure and the functions of cells and organisms. On the applied side of molecular biology, the development of recombinant technology requires a rationale for site-directed mutagenesis. This is fortunate for the field of protein structure prediction, which a decade ago was considered as rather esoteric and which now receives more attention and will thus, hopefully, progress more rapidly.

Some Characteristics of Proteins Related to their Structure Code

A polypeptide chain offers extreme diversity because of its chemical composition. For example, a peptide segment of only 20 amino acids, a common length for vaccine epitopes, has 20^{20} possible amino acid sequences of the 20 genetically coded amino acids, that is 1.04×10^{26} combinations or 173 times the Avogadro number. From this observation two conclusions can be drawn. First, we should not expect to obtain reliable statistical data from the few hundred proteins of known structure. Secondly, during the course of evolution, not all the possible combinations of amino acid sequences for each protein have been explored and this leaves plenty of room for innovative proteins.

Another characteristic of proteins is the relatively limited number of spatial structures which are observed, either at the level of secondary structures or at the level of supersecondary structures, i.e. the assembly of secondary structures in space. Two conclusions can be deduced. The first conclusion is that the structure code is degenerate *vis-à-vis* the amino acids, i.e. any amino acid can be observed in any conformation; only the frequencies of amino acids in a given conformation will vary. A recent compilation of these frequencies from Gibrat *et al.* [2] is presented in Table 1 which concerns about 11 000 amino acids when the amino acids are partitioned according to the classical definitions of α-helix, β-strand and coil or according to their grouping in various zones of the Ramachandran map (see definitions in legend of Table 1). We may note here that the two major zones of the map, h and e, are extensively

Table 1. *Distribution of amino acid residues according to their conformation*[1]

	Distribution of residues per torsional angle zones[2]					Distribution of residues per secondary structure[3]			
	h	e	g	x		H	E	C	Total
G	184	468	311	18	G	121	153	707	981
A	544	334	16	32	A	407	146	373	921
V	319	461	7	15	V	238	301	263	802
L	430	369	6	19	L	333	208	283	824
I	224	287	1	14	I	173	183	170	526
S	390	397	18	42	S	165	157	525	847
T	304	378	5	19	T	165	191	350	706
D	327	245	19	36	D	178	53	396	627
E	319	195	7	19	E	220	79	241	540
N	227	226	66	32	N	116	68	367	551
Q	216	183	8	12	Q	143	83	193	419
K	368	268	20	24	K	239	110	331	680
H	121	125	13	19	H	81	51	146	278
R	178	185	6	9	R	109	87	182	378
F	197	205	7	11	F	151	119	150	420
Y	162	229	13	15	Y	108	127	184	419
W	87	81	0	4	W	61	51	60	172
C	106	185	6	14	C	63	92	156	311
M	84	92	5	5	M	69	51	66	186
P	220	277	1	10	P	86	45	377	508

[1] Calculated from Gibrat *et al.* [2].

[2] h, e and g represent by in large the torsional angle zones of the Ramachandran map for right α-helix, ellipse centred at (−80°, −30°) with a major axis $a = 90°$ and a minor axis $b = 50°$ and the major axis tilted by 135° with respect to the ϕ axis; extended with $\phi < 0°$ and $\psi > 60°$ or $\psi < -140°$ or $\phi > 140°$ and $\psi < -140°$ or $\psi > 140°$ and left α-helix, ellipse centred at (80°, 20°) with $a = 90°$ and $b = 50°$ and the major axis tilted by 135° with respect to the ϕ axis, respectively. x represents all the other parts of the Ramachandran map with only 3.6% of the residues.

[3] Definition according to Kabsch & Sander's algorithm [9]. The number of residues in C is much greater than the sum of residues in g and x zones, then Coil includes residues in conformation h, α-helix or e, β-strand.

populated which means that the 'coil' conformation is by and large constituted of an irregular series of α-helix, h, and β-strand, e, conformations. The degeneracy of the structure code together with the over-determination of the spatial structure explains protein stability towards limited proteolysis or to the addition or deletion of peptide segments by genetic engineering at permissible points in the sequence [3]. The second conclusion is that only a limited number of motifs will have to be explored in molecular modelling, although still far too numerous for practical applications.

With only the amino acid sequence, no tertiary structure can be predicted in an acceptable range of accuracy, although some useful information can be obtained in some cases [4]. The major limitations come from the energy calculations involved in predicting tertiary structure. They are dependent both on the parameterization of the force fields, including solvent, and on the complexity of the energy minima which are not well sampled in the available computer times. The situation is different for the prediction of the secondary structure. A real success has been achieved and will be summarized here with some insights towards the protein-folding problem and the modelling of homologous proteins.

Some Characteristics of Prediction of Secondary Structure

Since the development of the pioneering predictive programs of Chou & Fasman [5], Lim [6] and Garnier *et al.* [7], there has been a slow, but continuous improvement in the accuracy of prediction which will be briefly discussed here.

All predictive methods have been developed and tested against protein structures determined by X-ray diffraction. Only structures determined to a certain degree of accuracy: 2.8 Å resolution with a refinement factor less than 0.25, are recommended to be used [8]; this corresponds to about 15 000 amino acid residues with the available X-ray data. The Kabsch & Sander algorithm [9] is regularly used to identify secondary structure from X-ray coordinates; this algorithm relies on the hydrogen-bonding network. It usually corresponds well with the crystallographic assignments of secondary structure, provided that the coordinates are known with the above accuracy.

An important feature of a predictive method is its degree of transferability from the known database to an unknown protein. This can be evaluated by measuring the probability index (the number of correctly predicted residues per predicted residues), either from a set of proteins removed from the database before the calculation of parameters or by computing the parameters for each predicted protein previously removed from the database (cross-validation). Cross-validation tests tend to reduce notably the probability index (expressed in percent) by a few points and particularly for neural network methods (see examples in Table 2).

We have also to bear in mind that all predictive methods are afflicted by a high standard deviation value of the probability index. For the COMBINE program of Biou *et al.* [10], the standard deviation is of the order of 7% for an average probability index of 65.5%. This means that for a three-state predic-

Table 2. *Comparison of secondary structure predictions*

	Without cross-validation		With cross-validation	
	3 states (%)	4 states (%)	3 states (%)	4 states (%)
Statistical methods				
Chou & Fasman [5]	51*		49*	
Garnier *et al.* [7]	57*	49	55*	
Gibrat *et al.* [11]	69.7		63.3	
Ptitsyn & Finkelstein [25, 26]			63	
Gibrat *et al.* [2]				62
Sequence similarity				
Nishikawa & Ooi [18]			60	
Sweet [19]			59	
Levin & Garnier [17]			63	
Pattern recognition				
Lim [6]	65*		56*	
Biou *et al.* [10]			59	
Qian & Sejnowski [24]			64.3	
Holley & Karplus [20]	68.5		63.2 (14 *P*)‡	
Combined prediction				
Biou *et al.* [10]			65.5 (67 *P*)‡	

* Kabsch & Sander [32].
‡ Number of proteins used for cross-validation.

tion, 95% of the predicted proteins will be predicted with an accuracy or probability index between 52% and 79%; a random prediction will be 38% of accuracy with a standard deviation of 0.5% [11]. One comment and one question can be made. The quality of a predictive method cannot be evaluated on a few proteins. An interesting question is why some proteins, or parts of proteins are so poorly predicted?

Recent Progress in Secondary Structure Prediction

The GOR method

The GOR method of Garnier, Osguthorpe & Robson [7] has been refined by introducing amino acid pair information and addition of dummy frequencies [11]. This brings the probability index of the method for a three-state prediction to 63.3% after cross-validation. New directional parameters have also been developed for α-helical, β-strand, turn and coil conformations [12], but single-residue information only still gives a less accurate prediction than when including pair information. However, these refined directional parameters should be useful for a rapid check of the effect on the local sequence of an amino acid replacement in protein design. A similar up-dating of Chou & Fasman parameters has also been undertaken [13, 14], with the addition of specific β-turn-type parameters [15].

The information theory approach has been recently revisited [2] to calculate conformational state probability for each residue.

According to the formalism described by Robson [16], the information value for the conformation, S, and a sequence of n amino acids, R, is:

$$I(S:\bar{S};\, R_1, \ldots, R_n) = I(S;\, R_1, \ldots, R_n) - I(\bar{S};\, R_1, \ldots, R_n) \tag{1}$$

where $\bar{S}$ represents the other possible conformational states.

By applying the definition of the information:

$$I(S;\, R_1, \ldots, R_n) = \log\,[P(S, R_1, \ldots, R_n)/P(R_1, \ldots, R_n) \cdot P(S)] \tag{2}$$

where $P(\ )$ are the probabilities of multiple events, $S, R_1, \ldots, R_n$, or simple event, S. It comes from eqns (1) and (2):

$$I(S:\bar{S};\, R_1, \ldots, R_n) = \log\,[P(S, R_1, \ldots, R_n)/P(\bar{S}, R_1, \ldots, R_n)] + \log\,[P(\bar{S})/P(S)] \tag{3}$$

The term $P(\bar{S})/P(S)$ can be considered as a normalization factor for the relative composition of the database in the conformations S and $\bar{S}$.

In the latest version of the GOR method, the right-hand side of equation (3) is approximated by pair interactions on 17 amino acid residues, 8 residues on each side of the residue to be predicted. Suppose I(App.) is the value of this approximation. Then, from eqn. (3):

$$e^{I(\text{App.})} = [P(S, R_1, \ldots, R_n)/P(\bar{S}, R_1, \ldots, R_n)][P(\bar{S})/P(S)] \tag{4}$$

with

$$P(S, R_1, \ldots, R_n) + P(\bar{S}, R_1, \ldots, R_n) = 1 \tag{5}$$

The two probability terms: $P(S, R_1, \ldots, R_n)$ and $P(\bar{S}, R_1, \ldots, R_n)$ can be calculated from eqns (4) and (5) and a prediction made from the highest probability value rather than the highest information value I(App.). This has been applied to the prediction of the Ramachandran zones, h, e, g and x with a probability index of 62% for four states [2]. This new approach has the advantage of calculating for each residue a probability for each of the four conformations, replacing the confidence scales previously developed. It also allows us to compare these calculated probabilities with those observed in the database (see Table 3). The agreement is relatively good. This finding has the important consequence that the local sequence of 17 amino acids is enough to account for the observed global distribution of residue conformations in the proteins of the database. This implies that the local sequence carries only, on average, about 65% of the constraints necessary for specifying the conformation of a residue in a protein of the database. The rest of the constraints are specified by long-range interactions, i.e. interactions between residues which are close in space, but distant along the sequence and specific for each protein, and which cannot yet be transferred to another protein except for homologous proteins, see below.

Predictions of secondary structure by sequence similarity

There have been several methods published whose algorithms are based on the assumption that short similar sequences of amino acids have similar secondary structure tendencies [8, 17–19]. They differ by the scoring of sequence

Table 3. *Observed versus calculated probability indices*

Consider a residue with the following calculated probability indices (PI): 0.54 for h, 0.27 for e, 0.15 for g and 0.02 for x. (See legend to Table 1 for definition of h, e, g, and x.) This residue is predicted h and lies in the PI interval 0.5–0.6 (column 4 in the Table). There is, theoretically, 54% chance to predict correctly this residue in zone h. The figures in the rows give the percentage of residues which being predicted in a zone with a given PI are effectively observed in this zone. In the example given above, the figure is 56%. The Table is reproduced from Gibrat *et al.* [2] with permission.

Calculated PI	0.2	0.3	0.4	0.5	0.6	0.7	0.8	0.9	1.0
% of h correctly predicted	*	36	51	56	67	73	80	93	
% of e correctly predicted	*	41	43	55	63	74	84	90	
% of g correctly predicted	*	32	35	53	54	61	*	†	
% of residues correctly predicted	*	38	48	55	64	73	82	91	
No. of residues involved	6	194	1570	3628	2804	1854	855	130	

* The numbers of residues involved are too small for the percentage to be meaningful.
† No residues are predicted for this state with this fraction.

similarity which has proved to be critical for the accuracy of the predictions. The similarity matrix of Levin *et al.* [8] seems to achieve the highest accuracy (see Table 2). The algorithms consist of an examination of the sequence similarity of a peptide length of n amino acids of the protein to be predicted against all peptides of the same length of all the proteins of the database. The value of n varies with the method from 11 by Nishikawa & Ooi [18], 12 by Sweet [19] to 17 by Levin & Garnier [17]. For the algorithm of Levin & Garnier, every time a peptide is found whose score calculated from the similarity matrix is less than a cut-off value of 7, the peptide is rejected. Every time the score is greater or equal to 7, its observed conformation is assigned to the tested peptide sequence with its similarity score. A simplified case is presented in Table 4. When every peptide fragment in the protein to predict has been compared, the secondary structure attributed to each residue is that which has the highest value after multiplication by a decision constant. As for the GOR method the optimal sequence length for prediction is 17 amino acid residues. A similar value was found by Holley & Karplus with a neural network approach [20]. The comparative accuracy of these methods are presented in Table 2. The similarity matrix plays a critical role for the accuracy of prediction, for instance the unitary matrix, the Dayhoff substitution matrix and the matrix recently developed by Risler *et al.* [21] have been compared and all give lower accuracy in secondary structure prediction than the similarity matrix of Levin *et al.* [8] (data not shown).

The accuracy of methods based on similarity is very sensitive to the presence in the database of homologous protein. Levin & Garnier [17] used this property to increase the accuracy of prediction of homologous proteins by weighting each similarity score according to the degree of homology of the reference protein. The results are presented in Table 5. For a set of homologous proteins, between 25 and 64% of identical residues, the probability index reaches 87.3% per polypeptide chain with a standard deviation of 6.6%. We will give below some uses of this property for amino acid sequence alignment.

Table 4. *The algorithm of Levin & Garnier* [*17*]

```
   - Principle:similar peptides of same sequence length n will have similar
               secondary structure tendencies.

   - Define    a similarity matrix of 20 x 20 Amino acids (Levin et al.,
 1986)
               a window length of n=17 amino acids
               a cutoff of 7

   - Simplified example case with  n=7, a cutoff of 2, a decision constant of
 1 and a unitary matrix

   - Unknown sequence:        S L C G T V A D L K ......

   - Reference data base         Seq   A L P G C T A L E G A V.......

                                 Obs   C C E E E E H H H H H H.....

   - Search:
                              A L P G C T A L E G A V....
        Step 1                S L C G T V A
                                                                Score=3

        Step 2                A L P G C T A L E G A V....
                                S L C G T V A
                                                                Score=1
        ....................all the data base

        Step i+1              A L P G C T A L E G A V.....
                              L C G T V A D
                                                                Score=0

        Step i+2              A L P G C T A L E G A V.....
                                L C G T V A D
                                                                Score=3
        ...................all the data base

        Step j+I              A L P G C T A L E G A V.......
                              C G T V A D L
                                                                Score=0
        ..................all the data base and all
                             the sequence to be predicted
```

- Match-score table and Prediction

Unknown	H	E	C	Pred.
S	0....	0.....	3.....	C
L	0.....	0.....	3+3....	C
C	0.....	3+3...	0.....	E
G	0....	3+3...	0......	E
T	0.....	3+3..	0......	E
V	0.....	3+3..	0.......	E
A	3+3....	0....	0......	H
D	3....	0.....	0......	H
:	:	:	:	:

Pattern recognition methods

The precursors to these methods were Schiffer & Edmunson [22] with the helical wheel rules followed by Lim [6]. The recent method we developed [10] identifies patterns of hydrophobic and hydrophilic residues along the amino

Table 5. *Secondary structure prediction of homologous proteins by the method of Levin & Garnier [17]*

Filename[1]	Identity[2] (%)	Basic prediction[3] (%)	Homologous prediction[4] (%)
4 APE	55	57.6	88.8
2 APP	55	56.4	89.5
2 ACT	48	53.7	83.5
2 ALP	39	50.0	77.8
2 ABX	42	81.1	75.7
5 CHA (1)	34–43	65.7	89.3
5 CHA (2)	41–45	60.8	87.6
3 CYT	44–59	57.3	99.0
1 CCR	40–59	61.3	91.9
3 C2C	40–44	64.3	87.5
2 EST	30–43	62.1	86.7
2 EBX	42	71.0	96.7
1 FDX	43	70.4	75.9
2 FDI	43	64.2	70.8
2 HHB (1)	35–45	61.7	92.9
2 HHB (2)	45	56.9	89.7
2 LHB	35	73.2	88.6
1 MCP (1)	47–64	67.3	83.2
1 MCP (2)	44	67.6	88.7
1 FB4 (2)	44	75.6	87.3
1 RE1	51–64	63.6	94.4
2 RHE	47–51	64.9	85.1
2 PKA (1)	34–42	63.8	96.3
2 PKA (2)	30–41	60.5	91.5
1 MBN	25	80.4	88.9
1 PPD	48	63.7	92.0
2 SGA	39	54.1	76.8
3 RP2	34–38	56.7	84.8
1 TPO	36–45	61.4	89.7
GLOBAL: 29	25–59	63.7 (S.D. 7.4%)	87.3 (S.D. 6.6%)

[1] Brookhaven databank filename.

[2] Several figures indicate that they are more than one homologous protein and the range of identical residues.

[3] Prediction made when all homologous proteins to the protein to be predicted (% of identical residues greater than 22%) are removed from the database.

[4] Homologous proteins are left in the database and the homologies are explicitly included as in Levin & Garnier [17].

acid sequence according to the known stereoregular structure of helices — one face of which is hydrophobic — and β-strands — all hydrophobic or alternate hydrophobic/hydrophilic residues. These simple patterns are surprisingly effective with a probability index of 59% for three states with a general tendency to be more accurate for small proteins. The search for more detailed patterns of amino acid sequences such as those undertaken by Rooman & Wodak [23] underlined the problem of extreme diversity of a polypeptide chain which prevents the finding of enough specific patterns for accurate secondary structure prediction.

Similarly the various neural network approaches [20,24] have not yet given a significant improvement in prediction accuracy (see Table 2). This also applies to the method of Ptitsyn & Finkelstein [25,26] which achieved a probability index of 63%. There is a good reason for this situation. Any method developed so far relies on a kind of statistical analysis of the same (small) sample of proteins and their convergence in accuracy confirmed our results of Table 3: with a limited database, it seems difficult to analyse more than the local effect of the sequence on the conformation.

Combining predictive methods

Combining several predictive methods has been shown to increase significantly the probability index. Biou *et al.* [10] have combined the prediction of the GOR [11], the program based on the algorithm of Levin & Garnier [17] or SIMPA for SIMilarity Peptide Analysis and a pattern recognition program or Bit Pattern from Biou *et al.* [10]. The probability index for a three-state prediction was 65.5%, the highest score reported so far after cross-validation and removal of the homologous proteins. The COMBINE program takes advantage of the difference between the highest and the second highest information values given by the GOR and/or the agreement between GOR and SIMPA or GOR and Bit Pattern which corresponds to 77% of the residues of the database. A second step of the COMBINE program extends the regular structures when they are interrupted by only one residue and there is a third step for a specific extension of the β-strands.

There have been attempts to combine experimental data, such as the secondary structure proportions from circular dichroism with the GOR [7] or a preliminary prediction of the family of proteins in terms of the proportions of secondary structure before the structure prediction [7,13,17,27,28]. The improvement of the GOR was found uncertain due to the significant error rate of circular dichroism data. Amino acid alignment of homologous proteins combined with the GOR has been found by Zvelebil *et al.* [29] to improve significantly the GOR prediction. Whether it would be the same with the new GOR version [11] or with COMBINE should be interesting to investigate.

Transferability of the accuracy to a new set of proteins

The answer to this question concerns all the users of the different predictive programs. Although cross-validation was simulated by removing the predicted protein from the set of proteins to calculate the parameters used for the prediction schemes, owing to the limitation of the database it was necessary to test the programs on a data set not used for their development. This was done for the programs GOR, SIMPA, Bit Pattern and COMBINE on two sets of proteins, one taken from a new set of proteins of the Brookhaven Protein databank determined by X-ray (Table 6) and a compiled data of protein structures determined by n.m.r. (Table 7). In both sets of proteins, the COMBINE method preserves its greater accuracy with no change for 'n.m.r.' proteins and a decrease for the 'X-ray' proteins with respect to the scores for the develop-

Table 6. *Prediction of proteins not included in the database and determined by X-ray diffraction*

Abbreviations: Su.C, Su.L, Su.M & Su.H are for subunits C, L, M and H, respectively.

		Method			
Protein name	Filename	GOR III (%)	SIMPA (%)	Bit Pattern (%)	COMBINE (%)
Adenylate kinase	3 ADK	60.8	71.1	56.2	64.9
Chymotrypsin inhibitor	2 CI2	61.5	44.6	58.5	56.9
Cytochrome *P*-450 CAM	2 CPP	57.8	61.0	57.8	64.2
Glutathione reductase	3 GRS	58.4	54.9	54.0	57.7
Phosphofructokinase	3 PFK	61.4	63.0	48.3	66.1
Photosynthetic reaction centre (Su.C)	1 PRC	64.2	66.3	69.3	75.0
Photosynthetic reaction centre (Su.L)	1 PRC	44.7	44.3	57.9	56.8
Photosynthetic reaction centre (Su.M)	1 PRC	58.2	50.5	58.2	63.2
Photosynthetic reaction centre (Su.H)	1 PRC	57.8	60.1	57.8	57.0
Pseudoazurin	2 PAZ	55.3	68.3	56.1	54.5
Troponin C	5 TNC	77.6	82.0	64.0	82.0
Trp repressor	2 WRT	71.2	76.0	44.2	75.0
Ubiquitin	1 UBQ	51.3	60.5	69.7	52.6
Mean %/residues		59.5	61.2	57.4	63.0
Mean %/protein		59.3	60.4	57.5	63.7
Mean deviation/protein		5.6	8.6	4.8	7.0
Total number of residues 3094					

ment set of proteins. One may notice that the Bit Pattern performed better on the 'n.m.r.' proteins as these are small proteins [12]. The decrease in accuracy for all the methods on the set of 'X-ray' proteins could reflect a lack of transferability, this set of proteins has a different global composition in secondary structures and also contains membrane proteins. However, the values of the standard deviations ranging from 5.6 to 8.6 are similar to the previous ones and prevent a real conclusion owing to the lack of statistical significance (at the confidence level of 99%) of differences in accuracy between these sets of proteins compared with the development set of proteins.

Role of Secondary Structure Prediction in Molecular Modelling by Homology

Prediction *per se* of secondary structure is of limited use. However, the choice of the mutation of an amino acid with respect to the conservation of the predicted secondary structure can be an aid in the choice of the amino acid. More useful is the aid for amino acid sequence alignment of homologous proteins. This is for two reasons. First, the modelling of tertiary structure is only feasible when the structure of a homologous protein is known and this

Table 7. *Prediction of proteins not included in the database and determined by two-dimensional n.m.r.*

EGF, epidermal growth factor.

	Method			
Protein name	GOR III (%)	SIMPA (%)	Bit Pattern (%)	COMBINE (%)
Histidine-containing protein (*Escherichia coli*)	56.5	57.6	58.8	61.2
Cytochrome *c N*-terminal helix	63.6	27.3	27.3	54.5
Secretin	63.0	74.1	48.1	61.5
Protein E-L 30 (*E. coli*)	51.7	51.7	51.7	60.3
Hirudin	70.8	67.7	66.2	75.4
Human ubiquitin	48.7	55.3	60.5	55.3
Cardiotoxin CTXIIb	60.0	63.3	71.7	66.9
Herpes simplex virus glycoprotein D-1 antigenic domain	59.1	63.6	45.5	59.1
Acyl carrier protein	76.6	66.2	59.7	75.3
α-Amylase inhibitor Tendamistat	64.9	71.6	48.6	68.9
Murine EGF	56.6	52.8	50.9	62.3
Lac repressor (*E. coli*) DNA-binding domain	54.9	62.7	64.7	62.9
Cardiotoxin VII	70.0	63.3	75.0	71.7
Proteinase inhibitor II A	51.8	51.8	60.0	48.2
Mean %/residues	61.9	62.3	60.0	65.5
Mean %/protein	60.9	59.8	56.1	64.2
Mean deviation/protein	6.5	8.8	10.4	8.3

relies on a correct amino acid alignment. Secondly, if the secondary structure prediction is sufficiently accurate, about 90%, it can be an aid for the amino acid alignment before the attribution of the coordinates of the known protein to the unknown one.

It is common knowledge that an amino acid alignment based only on the maximum percentage of identical residues does not correspond to the alignment deduced from the X-ray data. However, on the basis that the secondary structure is better conserved in homologous proteins, the alignment of the sequences can be checked by the alignment of the predicted secondary structure versus the observed ones. An example taken from the serine proteinase family is presented in Fig. 1, where the predicted secondary structures are compared with the crystallographic alignment taken from Craik *et al.* [30] and agree quite well. There have been many published alignment programs, and one of the most recent and most accurate has been developed by Rechid *et al.* [31]. When applied to α-lytic protease and proteinase A, the similarity to the crystallographic alignment is quite good, except in the regions of residues 28–93 (ALP numbering), where residues 45–80 are misaligned. If one aligns the three predicted β-strands of proteinase A with the corresponding ones in ALP, residues 49–58, 64–69 and 74–81 (see Fig. 2), 22 misalignments can be corrected leaving only 10 errors.

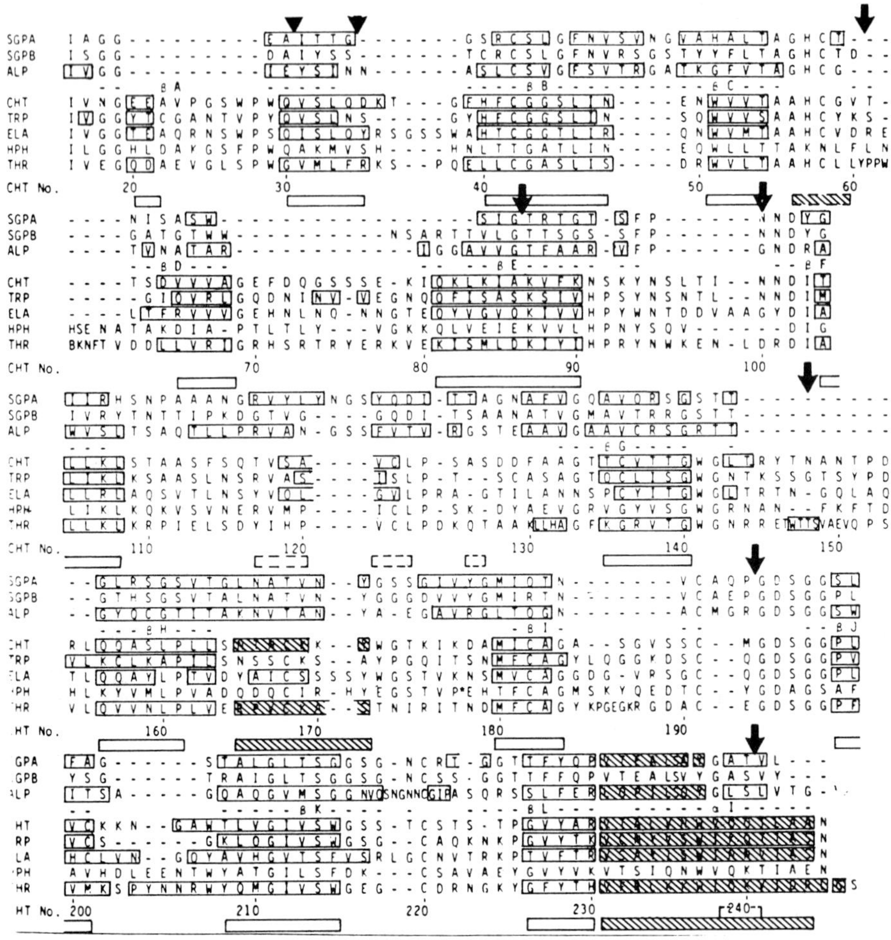

Fig. 1. *Comparison of crystallographic alignments and predicted secondary structures of the serine proteinases*

The crystallographic alignment is from Craik *et al.* [30]. On this alignment are superimposed the predicted α-helices (shaded boxes) and β-strands (open boxes). With the same symbols under the CHT numbering are the observed secondary structure for mammalian proteinases (continuous lines) and when they are different (with dashed boxes) the observed secondary structure of the fungus proteinases; ALP for α-lytic protease (E.C. 3.4.21.12), SGPA and SGPB for *Streptomyces griseus* protease A and B (E.C.3.4.21). The mammalian enzymes predicted are chymotrypsin B (CHT; E.C. 3.4.21.1), trypsin I (TRP; E.C.3.4.21.4), elastase II (ELA; E.C. 3.4.21.11) and thrombin (THR; E.C.3.4.21.5). The predictions were made by the SIMPA program from Levin & Garnier [17] with SGPA in the database for predicting ALP and ALP for predicting SGPA. For each mammalian enzyme the prediction was made with the other mammalian enzymes in the database. The homology was taken explicitly and the probability indices of the predictions are listed in Table 5, 4th column.

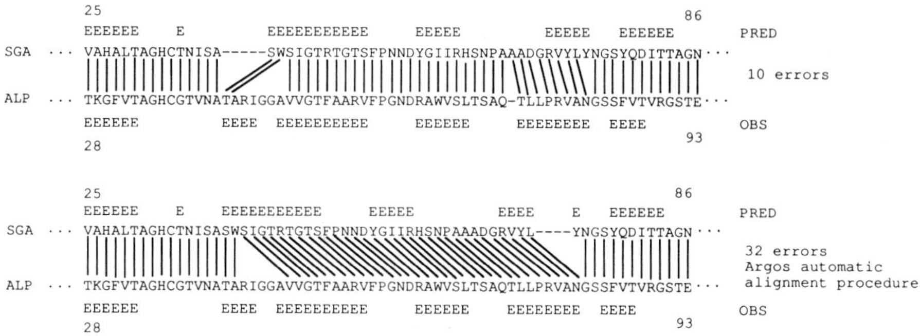

Fig. 2. *Amino acid alignment of S. griseus protease A (SGA) and α-lytic protease (ALP) with predicted secondary structure*

The amino acid alignment was done according to Rechid *et al.* [31] from a program generously provided by Pat Argos. Only the sequence alignment 28–93 is shown (bottom), this region being the only one with a discrepancy to the crystallographic alignment (bars) and the predicted secondary structure. OBS corresponds to the observed conformation of ALP and PRED to the predicted secondary structure of SGA with ALP in the database where E stands for β-strand and blanks for coil.

Conclusion

The success of secondary structure prediction methods which take account of only the local sequence, about 17 amino acids, reveals that part of the folding of a polypeptide is determined locally to about 65% on average. There is the possibility that the well-predicted regions of a polypeptide chain are the regions where the local structure is sufficiently stable and is not influenced by the long-range interactions and this would explain the observation that some short peptides of about 20 amino acids retain in solution the conformation they have in the protein they come from and thus behave as good artificial antigens. On the contrary, the less well-predicted regions would be those for which their final conformation is more strongly stabilized by long-range interactions. These peptides only would be in a different conformation in solution. If experimentally substantiated, the calculated probability indices per residue and per conformation should be useful for detecting artificial epitopes or to reinforce the stability of certain regions of a protein.

Finally, the development of predictive methods based on sequence similarity should be a useful tool in the modelling of homologous proteins.

Part of this work was financed by the E.E.C. Biotechnology Action Program, project 0149.

References

1. Goloubinoff, P., Christeller, J. T., Gatenby, A. A. & Lorimeer, G. H. (1989) *Nature (London)* **342**, 884–888
2. Gibrat, J. F., Robson, B. & Garnier, J. (1990) *Biochemistry* in the press
3. Charbit, A., Molla, A., Saurin, W. & Hofnung, M. (1988) *Gene* **70**, 181–189
4. Robson, B., Platt, E., Finn, P. W., Millard, P., Gibrat, J. F. & Garnier, J. (1985) *Int. J. Pept. Protein Res.* **25**, 1–8
5. Chou, P. Y. & Fasman, G. D. (1974) *Biochemistry* **13**, 222–245

6. Lim, V. I. (1974) *J. Mol. Biol.* **88**, 873–894
7. Garnier, J., Osguthorpe, D. J. & Robson, B. (1978) *J. Mol. Biol.* **120**, 97–120
8. Levin, J. M., Robson, B. & Garnier, J. (1986) *FEBS Lett.* **205**, 303–308
9. Kabsch, W. & Sander, C. (1983) *Biopolymers* **22**, 2577–2637
10. Biou, V., Gibrat, J. F., Levin, J. M., Robson, B. & Garnier, J. (1988) *Protein Eng.* **2**, 185–191
11. Gibrat, J. F., Garnier, J. & Robson, B. (1987) *J. Mol. Biol.* **198**, 425–443
12. Garnier, J. & Robson, B. (1989) in *Prediction of Protein Structure and the Principles of Protein Conformation* (Fasman, G. D., ed.), chap. 10, pp. 417–465, Plenum Press, New York
13. Chou, P. Y. (1989) in *Prediction of Protein Structure and the Principles of Protein Conformation* (Fasman, G. D., ed.), chap. 12, pp. 549–586, Plenum Press, New York
14. Prevelige, P. & Fasman, G. D. (1989) in *Prediction of Protein Structure and the Principles of Protein Conformation* (Fasman, G. D., ed.), chap. 9, pp. 391–416, Plenum Press, New York
15. Wilmot, C. M. & Thornton, J. M. (1988) *J. Mol. Biol.* **203**, 221–232
16. Robson, B. (1974) *Biochem. J.* **141**, 853–867
17. Levin, J. M. & Garnier, J. (1988) *Biochim. Biophys. Acta* **955**, 283–295
18. Nishikawa, K. & Ooi, T. (1986) *Biochim. Biophys. Acta* **871**, 45–54
19. Sweet, R. M. (1986) *Biopolymers* **25**, 1565–1577
20. Holley, H. W. & Karplus, M. (1989) *Proc. Natl. Acad. Sci. U.S.A.*, **86**, 152–156
21. Risler, J. L., Delorme, M. O., Delacroix, H. & Henaut, A. (1988) *J. Mol. Biol.* **204**, 1019–1029
22. Schiffer, M. & Edmundson, A. B. (1967) *Biophys. J.* **7**, 121–135
23. Rooman, M. J. & Wodak, S. (1988) *Nature* (*London*) **355**, 45–49
24. Qian, N. & Sejnowski, T. J. (1988) *J. Mol. Biol.* **202**, 865–884
25. Ptitsyn, O. B. & Finkelstein, A. V. (1983) *Biopolymers* **22**, 15–25
26. Ptitsyn, O. B. & Finkelstein, A. V. (1989) *Protein Eng.* **2**, 443–447
27. Busetta, B. & Hospital, M. (1982) *Biochim. Biophys. Acta* **701**, 111–118
28. Deleage, G. & Roux, B. (1989) in *Prediction of Protein Structure and the Principles of Protein Conformation* (Fasman, G. D., ed.), chap. 33, pp. 587–597, Plenum Press, New York
29. Zvelebil, M. J., Barton, G. J., Taylor, W. R. & Stenberg, M. J. E. (1987) *J. Mol. Biol.* **195**, 957–961
30. Craik, C. S., Rutter, W. S. & Fletterick, R. (1983) *Science* **220**, 1125–1129
31. Rechid, R., Vingron, M. & Argos, P. (1989) *Comput. Appl. Biosci.* **5**, 107–113
32. Kabsch, W. & Sander, C. (1983) *FEBS Lett.* **155**, 179–182

Biochem. Soc. Symp. **57**, 25–33
Printed in Great Britain

Inverting the Protein-Folding Problem

CHRIS SANDER

E.M.B.L., Postfach 10.2209, Meyerhofstrasse 1, D-6900 Heidelberg, F.R.G.

The Protein-Folding Problem

In the protein-folding problem, one would like to understand how a globular protein reaches its particular three-dimensional shape, starting from the linear sequence of amino acids coded on its gene. More precisely, given a protein sequence, one would like to be able to calculate (predict) the (average) atomic coordinates of the biologically functional protein. Methods from theoretical molecular physics, statistics, information and decision theory, artificial intelligence, etc. have over the last 20 years not solved this central problem of structural biology. Physical simulations at present take too much computer time or fail to correctly simulate the essential interactions. Statistical or empirical rules for structure prediction are generally weak. Three principal difficulties lie in the co-operativity of the folding process, in the delicate energetic balance between enthalpy and entropy and in the fact that protein sequences have evolved to satisfy diverse functional constraints and are not optimized merely to fold up correctly. Today, the protein-folding problem as a structure prediction problem remains fundamentally unsolved [1].

Partial solution to the problem: model building by homology

A partial solution of the problem is based on a remarkable fact of biological evolution: protein molecules retain their basic shape and some aspects of their sequence over millions of years, and across a wide spectrum of species. Whenever we can make the evolutionary connection between a protein of unknown and one of known three-dimensional structure (e.g. by comparing the information in their sequences), we can predict with fair accuracy the unknown protein structure. The art of model building by homology has developed into a powerful tool in the last 2 or 3 years and is the basis of most predictions of protein tertiary structure today (e.g. [2]). Exercise of the art requires skills in the alignment of multiple sequences and of three-dimensional structures, in computer graphics, in intelligent database access and in conformational molecular dynamics. In this context, the prediction of the effects of point mutations on protein structure and function is an area of intense current activity. However, the evolutionary approach to protein folding neither enables us to predict the structure of a new class of proteins nor to construct novel proteins which are unlike those found in nature.

Inverting the problem: protein design

For a more comprehensive solution of the problem, new approaches are needed. Several years ago, in an E.M.B.O. course in September 1986 [3], we attempted to break new ground by turning the question upside down 'Ask not to calculate the structure from the sequence, but ask instead to calculate a sequence from the structure'. In protein design, the problem to be solved is not structure prediction, but sequence prediction.

Inverting the problem leads to new ways of thinking about protein folding. Consider these new questions: 'Enumerate all protein sequences compatible with a given fold!'; or 'Design a sequence that will fold up like lysozyme!'; or 'How much sequence variation at the surface of a globular protein is compatible with maintaining the chain fold?'. The inverted problem may actually be easier to solve than the forward problem, for several reasons. First, the forward problem has only one essentially correct solution, namely the native structure. The inverse problem, if one is satisfied with one sequence that folds up as planned, has many solutions: we know that in nature a protein family typically has many different sequences of essentially identical three-dimensional structure. Secondly, sequence prediction needs pay no heed to functional requirements; the sequence can be optimized for the sole purpose of yielding the correct structure *in vitro*.

To illustrate the possible simplicity of a designed sequence compared with a natural sequence, consider the following example. Suppose you want to construct an all-α-helical protein with a certain given topology. You have at your disposal tables of single residue preferences for certain positions in secondary structures, in loops, in interfaces between secondary structures, and on the surface; also, you have tools to optimize packing by removing steric clashes. Suppose you then simple-mindedly choose for each position residues optimal merely from the point of view of single residue preferences, naively ignoring one of the principal rules of protein folding 'local sequence information is insufficient to fully determine the native fold' (e.g. the structure of oligopeptides as long as five or six residues can be dominated by the surrounding tertiary structure interactions [4]). Is it not conceivable that such a protein sequence, although clearly less complicated than natural sequences, may well do the job? Is it not conceivable that natural sequences are more complicated because they have been subjected to additional functional selective pressure and randomizing mutational events? We shall see in due course.

Protein design as sequence prediction is a new approach to the old problem of protein folding. To the extent that designed proteins are not subject to the same evolutionary constraints as are natural proteins, the protein-design problem transcends the classical protein-folding problem.

Designing Proteins *De Novo*

The number of proteins designed *de novo* is gradually increasing. An excellent recent review [5] presents important examples. Here, I merely mention proteins designed in the 1986 E.M.B.O. course [3]: two triose phosphate isomerase (TIM) barrels (4-fold symmetric $\alpha/\beta/\alpha/\beta$ barrels), two alternating α/β proteins (flavodoxin topology) and two α-helical bundles. The approach was to start from an idealized form of a well-known protein fold and to design one sequence that would produce that fold. This was done in two steps. (i) Choose a typical protein fold (say, a $\beta\alpha\beta$ nucleotide-binding fold) and idealize (simplify) its architecture. Incorporate, if possible, a binding site for a ligand the binding of which can be detected experimentally. The result is a backbone coordinate set that clearly belongs to a well-known structural class, yet is unlike any of the natural members of the class in detail. (ii) Invent, calculate and predict an appropriate amino acid sequence. The result is a full coordinate set (all atoms) of the designed protein and a list of its amino acid sequence. An attempt at experimental verification of one of these designs, the TIM barrel babarellin, is in progress, by expression of a synthetic gene (H. J. Fritz, personal communication).

Redesigning Protein Topology

A hybrid approach to the protein-design problem involves the redesign of a natural protein, keeping part of its original amino acid sequence and changing other parts, in the pursuit of particular design goals. In collaboration with S. Emery, W. Klaus & H. Blöcker at G.B.F. in Braunschweig, as well as M. Sagermann and D. Tsernoglou at E.M.B.L. in Heidelberg, and W. Eberle and P. Roesch at M.P.I.M.F. in Heidelberg, we have re-engineered the topology of loop connections in a bundle of four α-helices, the rop (repressor of primer) protein, and assayed the re-engineered structure by n.m.r. spectroscopy.

The particular hypothesis tested by the rop redesign was motivated by a number of fundamental observations about the changes in protein structures and sequences in natural evolution (e.g. [6]). The most strongly conserved aspects of the tertiary structure of globular proteins are a structural core and the topology of loop connections between elements of secondary structure in that core. In contrast, the detailed structure of loops undergoes large variations. The evidence comes from superposition of related natural protein structures. Corresponding multiple alignments of amino acid sequences show that evolutionary constraints on protein sequences are strongest in the interior of the conserved structural core, where residues make numerous intraprotein contacts, and in functionally important sequence positions. Sequence variation is much greater where residues make few contacts with other residues, e.g. in solvent-exposed surface or loop regions. The successful transfer, by cut-and-paste engineering, of a β-turn structure between two proteins, preserving the original loop structure, directly illustrates the possible variability and independence of loop conformation [7].

The fact that a structurally conserved core can be delineated in space and that there are stronger variations outside of that core lead to the 'core hypothesis': *the dominant factor determining the native protein fold is the interaction of residues in the structural core*; and the corollaries: (i) to first approximation, the effect on protein structure of sequence changes in the core and outside of the core, especially loops, can be separated, i.e. treated independently in prediction or construction of protein structure; (ii) radical changes in the topology of loop connections, within certain weak constraints, are consistent with the preservation of the structural core, although such changes are (almost) never observed in natural protein families. Proof or disproof of these statements about protein folding is very difficult to achieve in a single set of experiments. The rop redesign was intended as a first step in the verification (or eventual modification) of the 'core hypothesis'.

The natural protein on which the redesign was based is a small bacterial protein involved in regulating DNA replication via interacting with an RNA primer [8], called rop. This protein of 63 amino acids forms a homodimer, as shown by crystallography [9] and two-dimensional n.m.r [10]. Each subunit has an antiparallel pair of two helices and a flexible *C*-terminal tail of seven residues (57-GDDGENL-63). The two pairs of helices associate in antiparallel orientation to form a four-helix bundle (Fig. 1). Each of the four helices is oriented antiparallel to its two nearest neighbours and parallel to its diagonal, more distant, neighbour. Residues on the interior helix faces form a hydrophobic core. The loop forming the chain reversal of the helical hairpin involves only three residues on each subunit, L29, D30 and A31. At each end of the bundle there are four helix termini, two connected by a short loop, and two as polypeptide *N*- and *C*-termini.

The dimeric rop bundle has a definite handedness (left handed in our convention as well as in that of Presnell & Cohen [11]). To see the non-equivalence of the two chiral alternatives, follow the U shape of one subunit (helix–loop–helix) from *N*- to *C*-terminus; as you arrive at the *C*-terminus, the hydrophobic dimer interface (and hence the other subunit) lies either on one side or the other of the plane formed by the two helices (Fig. 1). To remember the left–right convention (valid for both monomeric and dimeric bundles), place the first two helices (counting from the *N*-terminus) in the plane of the paper, the other two behind that plane. The chain trace of the first two helices, from *N*- to *C*-terminus, runs clockwise from right-handed bundles and anti-clockwise for left-handed bundles. The existence of both right-handed and left-handed helix bundles in nature [11], suggests the possibility that the topology of loop connections can be arranged independently of helix packing.

Genetic alteration [12] has shown that the wild-type rop (Wtrop) structure is very resistant to single non-conservative point mutations on the helical surfaces and near the loop region. Evidence comes from functional integrity (interaction with RNA primer) or, in cases of functional change, successful overproduction and/or covalent modification of buried Cys residues. Deletion of five and insertion of two amino acids in the loop region also leaves the structure intact; conservative point mutations in the interior are tolerated, e.g. L48 to V and L41 to V, but at a loss of thermodynamic stability ([12]; G. Cesareni and M. Kokkinidis, personal communications). Wtrop is thus an excellent candidate for the type of re-engineering performed here. The analysis of known helix bundles and mutation experiments with rop suggested that the core hypothesis is valid for this class of proteins. The design of the left-handed monomer of rop (Lmrop) represents one extreme case consistent with this hypothesis: complete conservation of the protein sequence in the helical

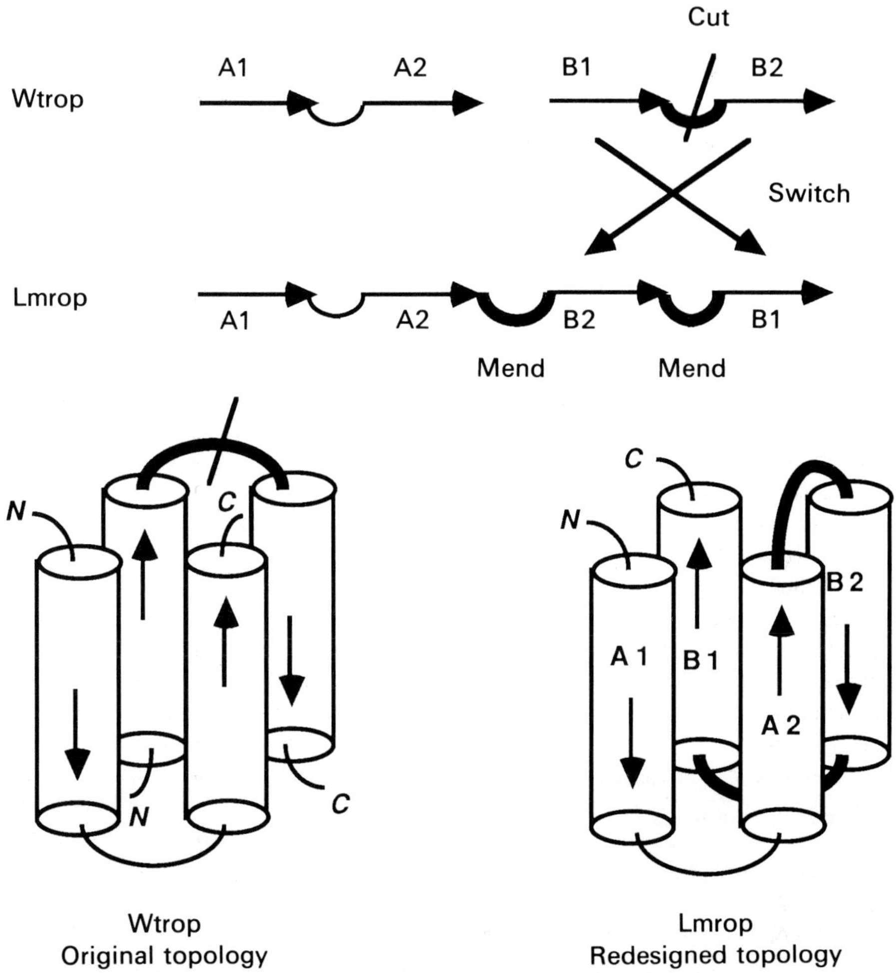

Fig. 1. *Cut-and-mend protein engineering: design steps from Wtrop dimer to Lmrop monomer involve deletion of one loop, addition of two new loops and a C-terminal extension*

Top: the orientation in space and packing of the four helices remains intact. Bottom: two helical segments are switched at the gene level. In detail: Wtrop has two identical subunits, say A and B, and a total of four helices, say A1, A2 and B1, B2 (cylinders/arrows). To achieve the desired topology of a left-handed four-helix bundle which starts with helix A1, delete the loop between helices B1 and B2; switch the order (in the chain sense) of these helices from B1, B2 to B2, B1); connect by a new loop helices A2, B2 and by another loop helices B2, B1; complete helix B1 by addition of a few residues. Loops involved in these changes are highlighted (thick lines).

regions, both in the interior (deemed essential) and on the helical surface (kept for convenience), and complete change in topology and amino acid sequence of loop regions (except on loop).

The redesign steps in going from the dimer Wtrop to the monomer Lmrop are these (Fig. 1): delete the loop on one two-helix subunit, say the second one; switch the order of the two helices in that subunit; and introduce two new loops, one between the subunits and one between the two helices of the second subunit, at the helix end opposite to where the original loop was deleted. The wild-type topology, A1–loop–A2–break–B1–loop–B2, becomes the Lmrop topology, A1–loop–A2–loop–B2–loop–B1–tail, where A1, A2, B1, B2 are α-helices.

Sequence design (Figs. 2 and 3), production of Lmrop proteins and determination of its three-dimensional structure were undertaken to verify the design and to provide a well-characterized vehicle for future protein folding studies [13]. Milligram amounts of pure protein allowed analysis by two-dimensional n.m.r. spectroscopy. Although we have not yet solved the structure of Lmrop in full atomic detail, the analysis proves that Lmrop does have four α-helical segments and that these helices come together in the three-dimensional structure in the same relative orientation, with the same interhelix contacts and the same bundle handedness at Wtrop.

The successful redesign of Lmrop is consistent with the core hypothesis, i.e. that the native fold is determined by interaction between the helices and that loops merely provide weak constraints which rule out certain arrangements. It remains an open question as to what precisely the constraints on loop connections are. We think that the principal restrictions are minimum loop length consistent with the arrangement of secondary structure elements in space and

helices A1+A2		loop L						helix B2	loop M			helix B1	tail N			
A1....A55	A56	L1	L2	L3	L4	L5	L6	B32...B56	M1	M2	M3	B3...B29	N1	N2	N3	name
S		K	K	P	G	Q	I		G	G	S		A	K	G	res

num	name	res	
	B1	M	deleted
	B2	T	deleted
	B30	D	deleted
	B31	A	deleted
55	A55	S	to avoid clash, replaces R A55
56	A56	F	kept
57	L1	K	continue helix after F A56
58	L2	K	continue helix
59	L3	P	turn PG as in loop PGQI of trypsin, contact with F A56
60	L4	G	as in trypsin res 174
61	L5	Q	as in trypsin res 175
62	L6	I	as in trypsin res 176
88	M1	G	flexible
89	M2	G	flexible
90	M3	S	possible H-bond to E B5
118	N1	A	for helix-helix packing - nothing bigger here !
119	N2	K	for charge interaction with COO of E's and/or C-term
120	N3	G	allows flexible adjustment of COO location

Fig. 2. *Protein sequence design of Lmrop and details of sequence changes relative to Wtrop* [*13*]

The amino acid sequence of Lmrop is derived from that of Wtrop by gene duplication (residue with names B1–B56 are a duplicate of A1–A56), deletion of four residues (B1, B2, B30, B31), mutation of one residue (A55), addition of nine residues in new loops (L1–L6 and M1–M3) and addition of a three-residue tail (N1–N3) for helix extension. 'num' is the sequential residue number in Lmrop (1–120); 'name' is the invariant residue name (chain and number, same in Wtrop and Lmrop when possible); 'res' is the amino acid type in one-letter code (A, Ala; G, Gly; etc.). The loop in trypsin was identified in Protein Data Bank data set 2PTN.

```
   M  T  K  Q  E  K  T  A  L  N  M  A  R  F  I  R  S  Q  T  L    A1
5' ATGACCAAACAGGAAAAAACCGCCCTTAACATGGCCCGCTTTATCAGATCTCAGACATTA

   T  L  L  E  K  L  N  E  L  D  A  D  E  Q  A  D  I  C  E  S    A1-loop-A2
   ACGCTTCTGGAGAAACTCAACGAGCTGGACGCGGATGAACAGGCAGACATATGTGAATCG

   L  H  D  H  A  D  E  L  Y  R  S  C  L  A  S* F  K* K* P* G*   A2-loop
   CTTCACGACCACGCTGATGAGCTTTACCGCAGCTGCCTGGCTAGCTTCAAAAAGCCGGGT

   Q* I* D  E  Q  A  D  I  C  E  S  L  H  D  H  A  D  E  L  Y    loop-B2
   CAGATCGATGAACAGGCAGACATCTGTGAATCGCTTCATGATCACGCTGATGAGCTTTAC

   R  S  C  L  A  R  F  G* G* S* K  Q  E  K  T  A  L  N  M  A    B2-loop-B1
   CGCAGCTGCCTGGCCAGGTTCGGTGGATCCAAACAGGAAAAAACCGCCCTTAACATGGCC

   R  F  I  R  S  Q  T  L  T  L  L  E  K  L  N  E  L  A* K* G*   B1-end
   CGCTTTATCAGAAGCCAGACATTAACGCTTCTAGAGAAACTCAACGAGCTCGCGAAAGGT 3'
```

Fig. 3. *DNA sequence design of the coding strand of the synthetic gene encoding Lmrop protein* [*13*] Residues in helical segments kept from Wtrop are underlined. Residues mutated or inserted (13 out of 120) are denoted by a star (*)

sufficient polar groups in side-chains to assure solvation. The stronger the loop constraints are, the easier the combinatorial problem in structure prediction; the weaker they are, the wider the options in protein engineering.

Support for the core hypothesis can also be derived from known cases of cyclically permuted protein sequences with preserved native structure. This simple form of topological rearrangement was observed *in vivo* comparing the structures of two leguminous lectins, favin and concanavalin A (Con A) [14,15]. Favin protein is similar to Con A in sequence (38% identical residues) and homologous in structure (both are the same type of antiparallel β-sheet sandwich), except for a cyclic permutation of the protein chain. The permutation is thought to occur not at the DNA level but after production of a precursor protein chain and without significant refolding of the initial translation product [16,17].

Pioneering experiments by Goldenberg & Creighton [18] and Kirschner and co-workers [19,20] produced two examples of proteins permuted cyclically by artificial means. The former made both a cyclic [21,22] and a cyclically permuted [18] form of pancreatic trypsin inhibitor by chemical modification and proved that the rearranged forms maintain the native fold. Luger *et al.* [19] achieved cyclic permutations of the yeast enzyme phosphoribosyl anthranilate isomerase (4-fold $\beta\alpha\beta\alpha$ barrel structure as deduced by sequence homology), by a gene construction permuting the order of the four $\beta\alpha\beta\alpha$ units from ABCD to DABC (plus shift by an extra helix in a second construct). Enzymic activity of the permuted forms as well as spectroscopic results (e.g. circular dichroism) indicated an essentially unperturbed structure compared with the wild type.

Preservation of globular structure in the face of a cut-and-mend operation appears consistent with the exon shuffling hypothesis [23–25]. However, we have no evidence that the individul helices of the bundle are independent folding units transferable to another three-dimensional structural context.

Rather, it seems likely that precise complementary packing of all four is required for proper folding. So, while sequential segment swap within one protein was shown here to be feasible, the transfer of folding units between proteins would be much more difficult, both for the protein engineer and for nature. Whether such transfer events are frequent in the natural evolution of globular proteins depends on the likelihood of mutations in the interface (in three-dimensions) between exchanged units, i.e. whether good core packing is achieved before the rearranged gene is eliminated.

Natural evolution of protein structures appears to have occurred along continuous pathways of point mutations and recombination of domains. A natural pathway from dimeric Wtrop to monomeric Lmrop would have to involve gene duplication, elimination of a stop codon, segment swap (invert the order of two helices) and mutations in loop regions, and all this without losing continuity of reading frame or viability of the organism. While there appear to be in nature examples of gene duplication resulting in a single protein of double size, it appears unlikely that natural evolution would be capable of evolving a molecule like Lmrop from a protein like Wtrop — the pathway from one to the other is very discontinuous. In this sense, protein engineering is capable of evolutionary jumps that transcend natural evolution.

The notion that solvent-exposed loop regions are the most strongly varying parts of protein structures in natural evolution is well established by the observation of numerous mutations in natural protein families. Our own and related [19, 20] results suggest that not only point mutations, and insertions or deletions, but also topological reconnections can be tolerated without affecting the core structure of a globular protein. The recognition that topological restrictions are much weaker than anticipated from the observation of naturally evolved protein structures may open up a new degree of freedom in the design of useful proteins.

It will be interesting to see if future repeat cycles of protein redesign, production, structure determination, analysis and redesign as performed once for Lmrop, will lend further support to the 'core hypothesis'. Meanwhile, it is already clear that the protein engineer has considerable latitude in the design of new structures.

References

1. Blundell, T. L., Sibanda, B. L., Sternberg, M. J. & Thornton, J. M. (1987) *Nature (London)* **326**, 347–352
2. Chothia, C., Lesk, A. M., Tramontano, A., Levitt, M., Smith-Gill, S. J., Air, G., Sheriff, S., Padlan, E. A., Davies, D., Tulip, W. R., Colman, P. M., Spinelli, S., Alzari, P. M. & Poljak, R. J. (1989) *Nature (London)* **342**, 877–883
3. Sander, C. (ed.) (1987) *Protein Design Exercises*, EMBL BIOcomputing Technical Document 1, available from EMBL, D-6900, Heidelberg, RAULFS@EMBL.bitnet
4. Kabsch, W. & Sander, C. (1984) *Proc. Natl. Acad. Sci. U.S.A.* **81**, 1075–1078
5. Richardson, J. S. & Richardson, D. C. (1989) *Trends Biochem. Sci.* **14**, 304–309
6. Chothia, C. & Lesk, A. M. (1986) *EMBO J.* **5**, 823–826
7. Hynes, T. R., Kautz, R. A., Goodman, M. A., Gill, J. F. & Fox, R. O. (1989) *Nature (London)* **339**, 73–76
8. Cesareni, G. & Banner, D. W. (1985) *Trends Biochem. Sci.* **10**, 303–306
9. Banner, D. W., Kokkinidis, M. & Tsernoglou, D. (1987) *J. Mol. Biol.* **196**, 657–675

10. Eberle, W., Klaus, W., Cesareni, G., Sander, C. & Rösch, P. (1990) *Biochemistry* **29**, 7402–7407
11. Presnell, S. R. & Cohen, F. E. (1989) *Proc. Natl. Acad. Sci. U.S.A.* **86**, 6592–6596
12. Castagnoli, L., Scarpa, M., Kokkinidis, M., Banner, D. W., Tsernoglou, D. & Cesareni, G. (1989) *EMBO J.* **8**, 621–629
13. Emergy, S. C., Sagermann, M., Eberle, W., Klaus, W., Blöcker, H., Roesch, P., Tsernoglou, D. & Sander, C. (1990) submitted
14. Reeke, G. N. Jr, Becker, J. W. & Edelman, G. M. (1975) *J. Biol. Chem.* **250**, 1525–1547
15. Reeke, G. N. Jr & Becker, J. W. (1986) *Science* **234**, 1108–1111
16. Hemperly, J. J., Mostov, K. E. & Cunningham, B. A. (1982) *J. Biol. Chem.* **257**, 7903–7909
17. Bowles, D. J. *et al.* (1986) *J. Cell Biol.* **102**, 1284–1297
18. Goldenberg, D. P. & Creighton, T. E. (1983) *J. Mol. Biol.* **165**, 407–413
19. Luger, K., Hommel, U., Herold, M., Hofsteenge, J. & Kirschner, K. (1989) *Science* **243**, 206–210
20. Goldenberg, D. P. (1989) *Protein Eng.* **2**, 493–495
21. Chazin, W. J., Goldenberg, D. P., Creighton, T. E. & Wuethrich, K. (1985) *Eur. J. Biochem.* **152**, 429–437
22. Goldenberg, D. P. & Creighton, T. E. (1984) *J. Mol. Biol.* **179**, 527–545
23. Gilbert, W. (1978) *Nature (London)* **271**, 501
24. Doolittle, W. F. (1978) *Nature (London)* **272**, 581–582
25. Blake, C. C. F. (1979) *Nature (London)* **277**, 598

Biochem. Soc. Symp. **57**, 35–48
Printed in Great Britain

Structural Homology in Ligand-Specific Transport Proteins

A. C. T. NORTH

Astbury Department of Biophysics, University of Leeds, Leeds LS2 9JT, U.K.

β-Lactoglobulin (BLG) is the most abundant protein in the soluble (whey) fraction of cows' milk and that of other ruminants. It is a very stable and well-characterized protein, but its function has not yet been unequivocally identified. Although it was first crystallized in the 1930s, its three-dimensional structure was not determined to sufficiently high resolution to elucidate its chain fold until 1985 [1]. The crystallographic studies have been carried out by the groups of Sawyer in Edinburgh (using crystals grown on the high pH side of the well-known Tanford transition) and North (with crystals grown on the acid side of the transition). The differences between the two forms appear to be comparatively slight and will not be discussed here.

The cow BLG is a dimeric molecule comprising identical monomers with 162 amino acid residues. The polypeptide chain is largely in the β-conformation, with a total of nine strands and one helix. The view shown in Fig. 1 is obtained looking along the two-fold rotation axis that relates the two monomers in the dimer. In Fig. 2, the dimer has been rotated so that different aspects of the two monomers can be seen, and these reveal noteworthy characteristics of the tertiary structure. In the right-hand view the chains can be seen to be arranged in the form of two sheets, the axial directions of which run at right angles to each other. In fact, eight of the strands form a calyx-shaped flattened barrel, with the strands spiralling around the barrel. The topology of the arrangement is very simple, with alternate strands running up and down with respect to the axis of the barrel and neighbouring strands being formed from adjacent segments of the polypeptide chain. Strands ABCDEF constitute a sheet forming one side of the barrel, while strands FGH and A form the other side. Strands A and F are thus common to both sheets. The single helix follows strand H and is itself followed by strand I, the *C*-terminal part of the chain; as can be seen from Fig. 1, the two I strands in a dimer run anti-parallel and provide the principal interactions between the two monomers. The BLG chain fold is different in topology from a number of β-proteins the structures of which had previously been determined (such as the trypsin family and the immunoglobulin domains) in which adjacent β-strands are not always from consecutive parts of the polypeptide chain. The left-hand monomer in Fig. 2 represents a view looking along the axis of the barrel and into a deep calyx-shaped pocket.

Almost simultaneously with the determination of the BLG molecular struc-

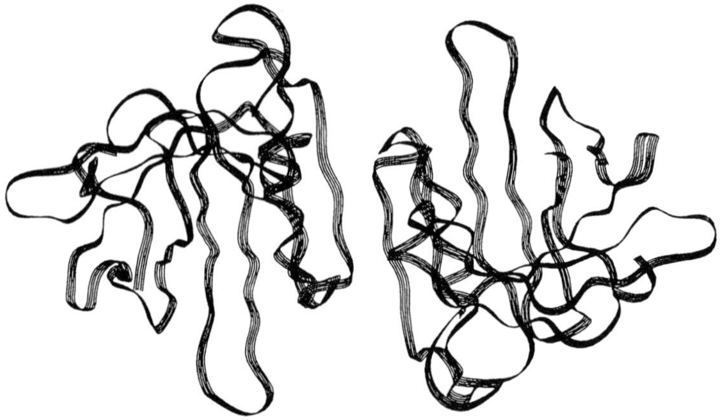

Fig. 1. *Ribbon diagram showing the chain fold of the BLG dimer, as seen down the axis of 2-fold symmetry*

ture, Jones and his colleagues obtained the crystal structure [2] of human serum retinol-binding protein (RBP) and it was immediately apparent that BLG and RBP have closely similar three-dimensional structures [1,3]. Fig. 3 shows a superposition of the two structures from which it can be seen that they share a common framework provided by the β-strands and helix, but they differ in the loops that connect together these secondary structure elements. Although Pervaiz & Brew [4] had already noted some sequence similarities between RBP and BLG, the degree of homology is not high: when the sequences are aligned with respect to the three-dimensional structures (i.e. with insertions and deletions occurring in loop regions) only 35 of the residues are identical; nevertheless, the three-dimensional structures are closely similar, with 129 out of the 162 residues of BLG having an r.m.s. difference in position

Fig. 2. *The BLG dimer rotated around a horizontal axis*

The left-hand monomer shows the view into the binding pocket and the right-hand monomer shows how the strands of the near and far side of the pocket run in perpendicular directions.

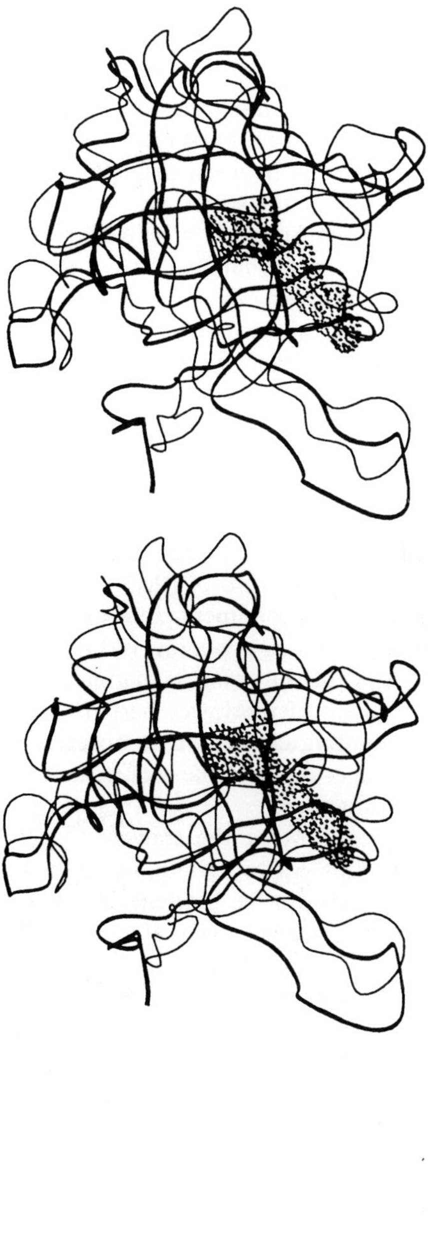

Fig. 3. *Stereo diagram showing the superposition of the main-chain of BLG (thick lines) on that of RBP (thin lines)* The retinol molecule, as bound in RBP, is shown by the dotted surface.

of less than 3 Å. The RBP structure was in fact of a complex with retinol bound, the pigment molecule being located in the pocket within the calyx. The similarity between the two proteins supported the proposal, based on studies of the binding to BLG of a range of hydrophobic molecules, that BLG also might be a transport protein having the function of carrying retinol in milk from the cow to her calf. Molecular modelling shows that retinol can readily be accommodated within the BLG pocket in the corresponding position to that adopted in RBP, but in fact the pocket in BLG is substantially deeper than the RBP pocket because of the substitution of three out of the five aromatic residues lining the RBP pocket by smaller side-chains in BLG. By model building, we found that the retinol molecule could readily be moved further into the BLG pocket (Fig. 4), and it could be located in a position that allowed the formation of a hydrogen bond between the retinol hydroxyl group and a polar side-chain. This location is consistent with spectroscopic observations that the retinol molecule is bound to BLG in a hydrophobic environment and also that the fluorescence of a tryptophan residue is affected when the ligand is bound (a tryptophan residue is among those lining the BLG pocket). Although this model is therefore attractive, the only crystallographic study so far reported of a BLG/retinol complex [5] shows the retinol molecule bound to a crevice on the surface of the protein and not in the internal site corresponding to that occupied by retinol in RPB. It remains to be established whether this represents an artefactual mode of binding, as the observations on other related proteins to be discussed below all show ligands bound within the central cavity, as in RBP and our model.

The fact that BLG will bind retinol does not of itself demonstrate that retinol transport is the primary function of BLG *in vivo*, but support for this being the case has come through experiments by Findlay, Sivaprasadarao and co-workers [3], showing that a complex of BLG and radioactive-iodine-labelled retinol binds in a pH-dependent manner to purified microvilli obtained from a specific region of the lower part of the small intestine of a young calf. Such specific binding could not be observed with a preparation from a 6-month-old calf, suggesting that the requirement for uptake of retinol by this route had ceased by that age.

While it is therefore plausible that the transport of retinol is indeed an important function of BLG, it is not yet clearly established that this is the only function of the protein and, indeed, the binding pocket is sufficiently deep to accommodate rather larger molecules than retinol.

The three-dimensional structures of further examples of this superfamily of proteins have subsequently been determined. These include the bilin-binding protein from *Pieris brassicae* [6,7] and the related insecticyanin from the tobacco hornworm [8] which exhibit the same structural framework of β-strands with interconnecting loops, forming a ligand-binding pocket. The bilin molecule seen in the crystal structure of both of these proteins is not, as might have been expected, in an extended form, but is coiled into the form of a flat helix. The binding pocket is thus fairly wide in order to accommodate the ligand. Although the bilin-binding molecules look somewhat different from BLG and RBP, this is largely because the interconnecting loops are rather longer; the β-strands of the framework superimpose quite closely [15]. The bilin-binding proteins are responsible for the coloration of the insects in which they occur, and they presumably serve the function of camouflaging them.

It has now been found that a related family of proteins is based upon a similar general structural principle to BLG, but involving a barrel with ten strands instead of eight. The first such example to have its structure determined in three-dimensions was the P2 myelin protein (which binds fatty acid molecules) [9] and this has been followed by the rat intestinal fatty acid binding protein (FABP) [10] and the photoactive yellow protein from *Ectothiorhodospira halophila* [11] which have a similar structural framework.

The *N*-terminal regions of these molecules all show homology to the BLG/RBP family, but this homology is not retained beyond the end of the first strand of the structure. It is in fact possible to superimpose part of the barrel, but the insertion of the additional strands prevents more than a local superposition. The fatty acid has been clearly located in FABP, in which it can be seen that the polar groups of the fatty acid occur in the middle of the binding pocket, where they interact with amino-acid side-chains. This is in contrast with the retinol/RBP complex, in which the polar group of the retinol lies at the surface of the protein.

It is believed that the cellular RBP and retinoic acid-binding proteins belong to this ten-stranded family rather than the eight-stranded one of which serum RBP is a member. One further example of this superfamily is the protein streptavidin [12] which appears to have the eight-stranded architecture, although its sequence is apparently more closely homologous to that of FABP than to BLG or RBP.

Comparison of amino acid sequences has revealed a number of other proteins that are homologous to BLG [13,14]. Typically, the monomeric chains of these proteins are 160–190 amino acids long and, when compared pair-wise, they show approximately 20% identity. They share two regions of above-average homology, each approximately 15 residues long, one centred on residue 17 (BLG numbering) and the other centred on residue 99. The most conspicuous features of these two regions are the tripeptide GXW in the first region and the hexapeptide TDYXXO, where O represents an aromatic side-chain, in the second. There is some variation in the latter group, but G is almost universal and W invariant in the first group. In addition, a disulphide bridge is formed between a residue near the *C*-terminus of the molecule and either residue 66 or residue 36, both alternatives being compatible with the same overall chain fold. These sequence similarities have led to the suggestion [13,15] that the molecules in this superfamily share a common chain fold designed to facilitate the transport of hydrophobic ligands. Some of the proteins concerned are listed in Table 1, together with the relevant ligands where these are known. In some cases, the protein appears to be concerned with transport, either within or between organisms. In such cases, it is to be presumed that the ligand-binding protein is specific for a cell receptor to or from which the ligand is to be transported. In other cases, such as the bilin-binding proteins, it is not clear whether the role of the protein is purely a static one such as conferring an appropriate colour on the organism.

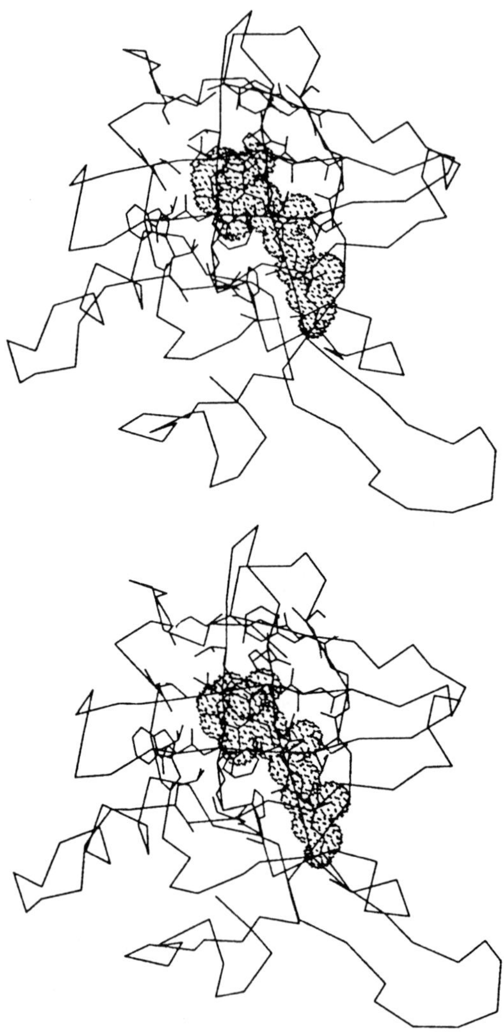

Fig. 4. *Stereo diagram showing the minimum-energy position of retinol (dotted surface) within the BLG molecule (shown mostly as an α-carbon chain plot, with all atoms of those residues lining the binding pocket)*

Table 1. *Properties and functions of the ligand-binding protein family*

Protein	Size (residues)	Source	Ligands	Properties
BLG	2 × 162	Milk whey	Retinol and other hydrophobic molecules	Vitamin transport to gut receptor
Serum RBP	182	Serum	Retinol in complex with transthyretin	Retinol transport to the eye
Purpurin	196	Retina	Retinol, heparin	Cell survival
α-1-Acid-glycoprotein	187	Hepatic cells, serum, urine	Progesterone	Mediates inflammatory response
Androgen-dependent secretory protein	165	Epididymal luminal fluid	?	Binds to sperm membrane
Endometrial α-2-globulin	~21 kDa	Amniotic fluid	?	Synthesized in early pregnancy
α-1-Microglobulin (Protein HC)	183	Cerebrospinal fluid, serum	IgA and a yellow-brown retinoid	Mediation of neutrophil chemotaxis
Apolipoprotein D	169	Gland and gut secretions	Lecithin, cholesterol	Lipid transport
Major urinary protein (α-2-microglobulin)	162	Hepatic cells, serum, urine	Pheromones?	Signalling?
Bowman's gland odorant-binding protein	160	Nasal mucus	Odorants	Presentation of odorants to receptor?
Insecticyanin (bilin-binding protein)	189	Haemolymph	Biliverdin IX	Camouflage
Crustacyanin	170 + 177	Carapace	Astaxanthin	Coloration, photo-reception

We have investigated two of these proteins in some detail. The first of these is the protein crustacyanin, one of a family of carotenoid-binding proteins found in the carapace of crustacea, where they are responsible for the animal's coloration. In a number of these proteins, the pigment responsible for the coloration is astaxanthin. The absorption maximum characteristic of free astaxanthin is shifted by up to more than 100 nm as a result of interaction with the protein moiety in the various complexes. In the case of the lobster carapace, the resultant colour is blue (with the colour changing to red when the lobster is cooked). While the protein/pigment complex may simply serve the function of coloration, it might also be a primitive photoreceptor system and it is of interest that the spectral shift that occurs when the pigment is bound to the protein is a phenomenon of relevance to our understanding of other protein/pigment systems involved in photosynthesis, photoprotection and vision.

Crustacyanin can be prepared in various molecular forms [16,17], of which β-crustacyanin is a dimer comprising two chains, A and C, the association between which is stabilized by the binding of two molecules of astaxanthin. The affinity of astaxanthin for each monomer is lower than its affinity for the dimer and carotenoids having longer or shorter chains than astaxanthin have lower affinities than astaxanthin itself. The amino acid sequences have been determined for both the A and C chains [17]. There is a high degree of homology between the two chains, with 38% identity and all five cysteine residues being conserved. The A chain has been modelled by homology with RPB, an optimum alignment showing 33 identities (cf. the 35 identities between BLG and RBP). Secondary structure prediction and hydrophobic moment analysis suggested the presence of eight strands in corresponding positions to those in RBP, and the model has been constructed with the β-strand framework of RBP. The connecting loops have been built using standard stereochemical parameters, and the resultant conformation has been subjected to a limited degree of energy refinement.

We have constructed a model for the C chain by homology with that for the A chain, the chain itself showing a lower degree of homology with RBP. The disulphide bridge arrangement is similar to that found in insecticyanin and bilin-binding protein, rather than the BLG/RBP arrangement.

The models are consistent with accepted protein structural principles, such as buried groups being hydrophobic, but we do not claim more than that they support the proposal that the crustacyanin chains share a common fold with the BLG family.

Although the binding pocket is quite deep, it is not sufficiently long to accommodate entirely an unfolded molecule of astaxanthin. Spectroscopic studies of the astaxanthin/crustacyanin complex are consistent with the pigment molecule having an extended form and we propose that, while the major part of the pigment molecule is bound within the pocket of one subunit of the dimer, the remaining part is located in a depression in the surface of the opposing monomer. This would be consistent with the binding studies that show that ligand binding contributes to the formation of the dimer. We cannot give a detailed explanation for the cause of the spectral shift observed when astaxanthin binds to the protein, though we note that side-chains lining the binding pocket are close to the polyene chain of the pigment and might twist it from its normal all-*trans* conformation. Deep within the binding cavity lie a threonine and a tyrosine side-chain, which could form hydrogen bonds to the keto and hydroxy groups of the astaxanthin. Spectroscopic studies of the binding of astaxanthin analogues show quenching of the fluorescence due to tryptophan residues of the protein; the invariant tryptophan of the ligand-binding superfamily is the only such residue in the A chain and one of the two such residues in the C chain; it projects into the binding pocket where it would undoubtedly be affected by the presence of ligand. Fig. 5 shows our model of the astaxanthin/C chain complex and Fig. 6 shows a model for the AC dimer with two pigment molecules.

We know that the relative disposition of the two monomers is not correct in detail; cross-linking experiments are in progress with a view to identifying residues of the two monomers that are near to each other (P. F. Zagalsky, personal communication). We have obtained small crystals of crustacyanin, but have not yet obtained sufficiently large ones for a full structural analysis.

Particularly interesting examples of the ligand-binding protein family are those that are involved in the binding of odorant molecules. These include on the one hand the urinary proteins that are presumably designed to facilitate the transmission end of a biological signalling system, and on the other hand proteins found in the nasal mucosa, which are presumably concerned with the reception end of pheromone signalling. The latter class are not the actual odorant receptors, but they are assumed to be concerned with conveying the odorant molecules through the nasal mucosa to the receptors. We have carried out molecular modelling studies of the cow nasal mucosa protein, using BLG as the starting point in this case as odorant-binding protein (OBP) is more nearly homologous to BLG than to RBP. We have used a similar procedure to that used for crustacyanin, constructing the β-sheet framework as in BLG, and then adding the appropriate OBP loops. Again, the loops have been subjected to limited stereochemical refinement. As with crustacyanin, the amino acid substitutions conform to the normal protein structural principles. OBP is known to bind a range of odorant molecules with varying degrees of affinity. These include linear, cyclic and bicyclic molecules, geosmin (the odorant associated with 'earthy' smell), a bicyclic compound, showing one of the highest affinities for the protein. The basis of the specificities and relative affinities is not yet clear, but obviously the ligand-binding pocket must be appropriate to accommodate a range of different shapes. In our model, the pocket is indeed able to accommodate the various odorants known to bind to OBP, and Fig. 7 shows its complex with geosmin. We are at present carrying out crystallographic studies of the related, but not identical, rat urinary protein.

It is clear that many members of this superfamily must have dual molecular recognition properties, for the ligand to be transported and for a cell receptor to which the ligand is to be delivered or from which it is to be collected. As we have previously pointed out [20,21], it is interesting that the highly homologous regions of the sequences seem to occur mainly on the external surface of the molecule, as seen in Fig. 8, which shows their positions in BLG. It does

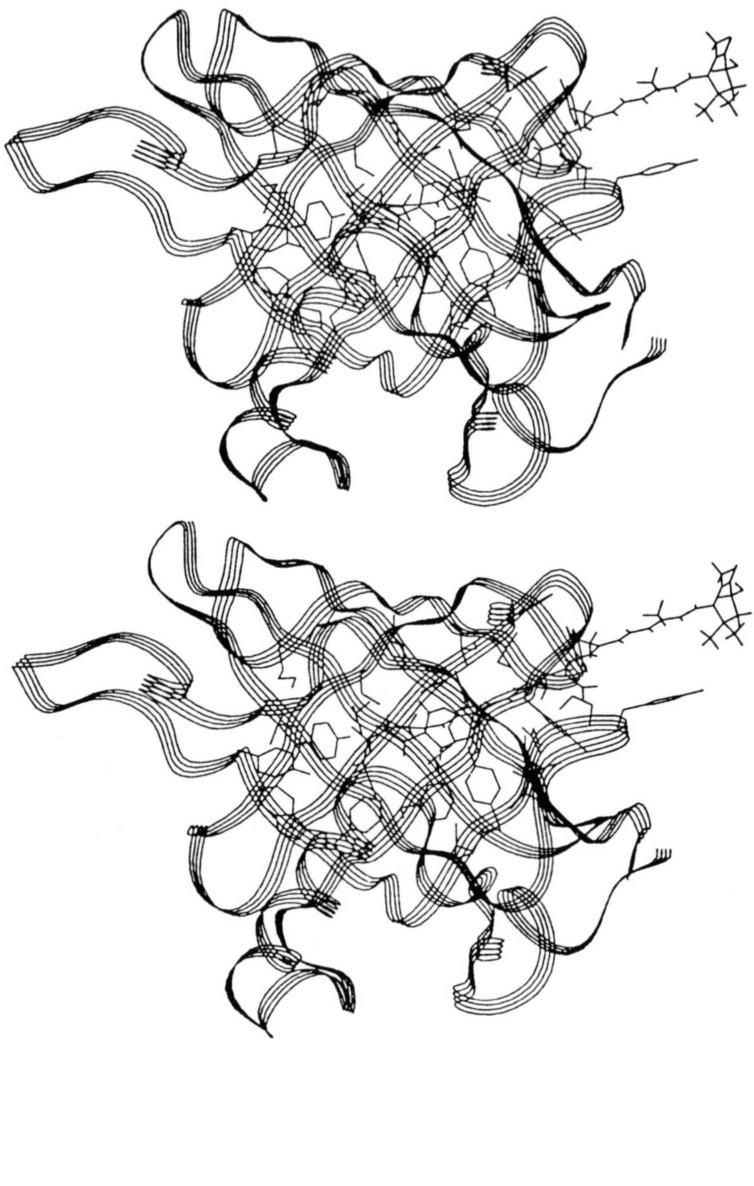

Fig. 5. *Stereo diagram of model of a crustacyanin C chain (ribbon, with some side-chains added) complexed with astaxanthin, which protrudes from the monomer*

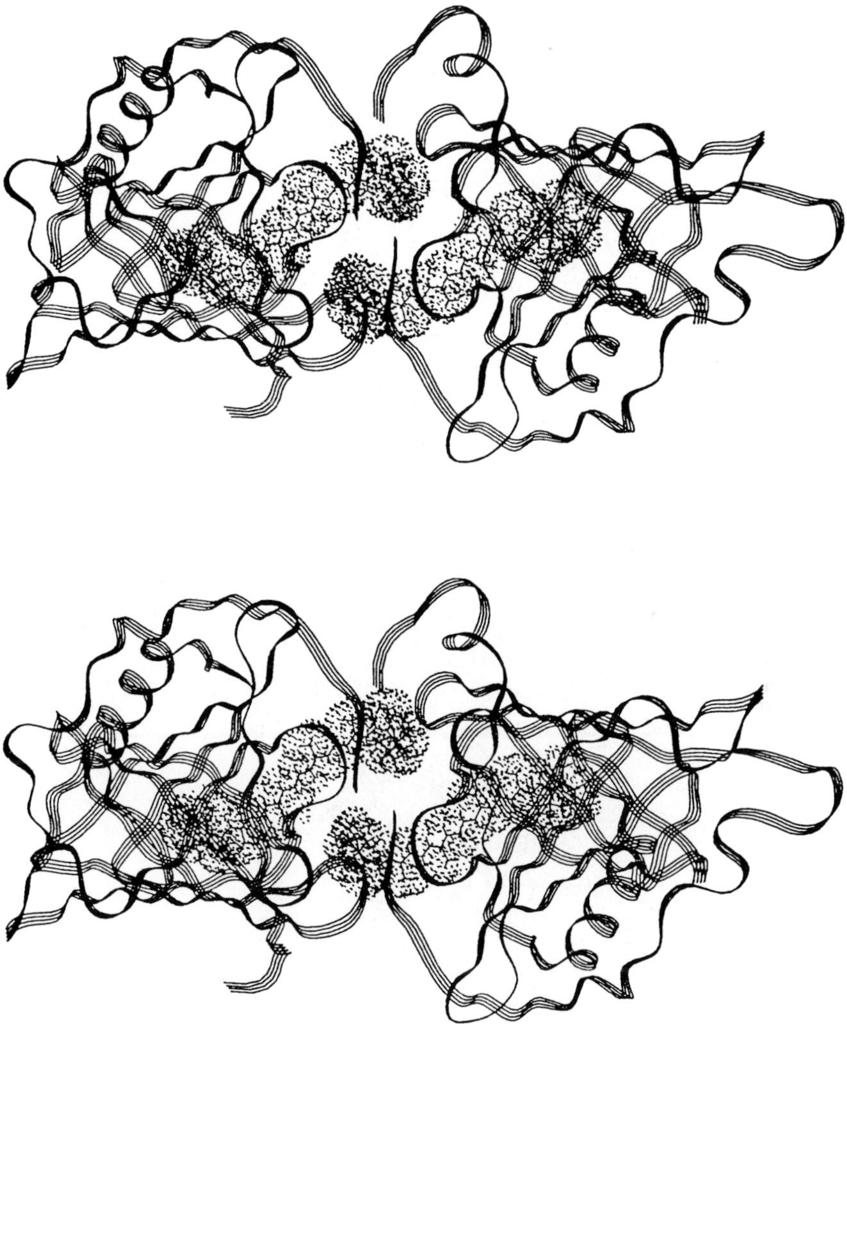

Fig. 6. *Stereo diagram of a model of a crustacyanin dimer with two bound astaxanthin molecules (dotted surfaces)*
Each astaxanthin molecule is mainly within the pocket of a crustacyanin chain, but it projects into a depression in the other protein monomer.

Fig. 7. *Stereo diagram of model of the cow nasal mucosa odorant-binding protein, with geosmin bound (dotted surface)*

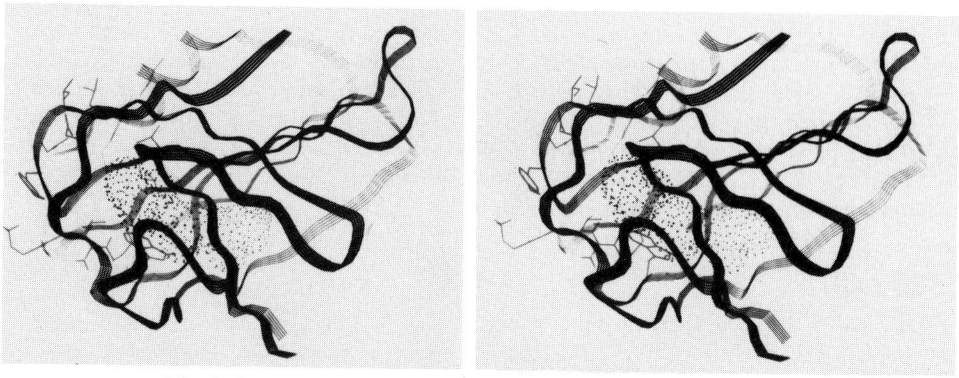

Fig. 8. *Stereo diagram of the BLG monomer with bound retinol (dotted)*
The patch of invariant or highly homologous residues is at its left-hand end.

not appear that they have a direct role in ligand-binding, nor do they have an obvious role in stabilizing the three-dimensional structure of the proteins. We therefore suggest that they might be concerned with binding to receptor cells, on the presumption that receptor sites with different affinities may nevertheless share common structural features, as has been demonstrated for a number of receptor cell systems. The variable residues on the surface of the ligand transport proteins presumably interact specifically with complementary discriminating features of the receptor cells. As has already been pointed out, the one completely invariant residue in the whole superfamily appears to be the tryptophan that points into the ligand-binding pocket; at the present time, one can only speculate as to its role, but a possibility is that it is involved in some kind of triggering mechanism that is sensitive to whether the pocket is empty or full. Sivaprasadarao & Findlay [22] have shown that molecules of the RBP/retinol complex are able to displace ligand-free RBP molecules from their specific cell receptors, thus suggesting that some ligand-sensitive mechanism must exist to control the affinity of the protein for the receptor.

Doubtless other examples of this ligand-specific protein family will come to light once it has been clearly established what features constitute the essential characteristics of this protein fold. We have emphasized the characteristic occurrence of two short sequence motifs, a GXW triplet followed some 70 residues later by TD with an aromatic residue shortly afterwards.

This combination of motifs occurs in the retinoic acid receptor protein (see, for example, [23]), a member of the family of receptor proteins that includes the thyroid hormone receptor. These receptor proteins are characterized by a zinc-finger domain towards their *N*-terminus, together with one or more ligand-specific domains. While the occurrence of a pair of tripeptides a defined distance apart might be due to chance, the fact that the receptor is specific for retinoic acid argues in favour of the occurrence of this characteristic feature of retinol-binding proteins being significant. We have applied secondary structure prediction methods to the ligand-binding domains of the retinoic acid receptor proteins, but they do not appear to give conclusive support for the presence of the BLG/RBP chain fold.

Many collaborators have contributed towards the work described in this review and I wish to thank Drs E. E. Eliopoulos, S. J. Yewdall and R. Cooper (Department of

Biophysics, University of Leeds), J. B. C. Findlay and A. Sivaprasadarao (Department of Biochemistry, University of Leeds), L. Sawyer and M. Z. Papiz (Department of Biochemistry, University of Edinburgh), T. A. Jones, M. E. Newcomer and P. J. Kraulis (Department of Molecular Biology, University of Uppsala), P. F. Zagalsky (Department of Biochemistry, Royal Holloway Bedford New College) and A. Cavaggioni and R. Tirindelli (Institute of Human Physiology, University of Parma). I also thank the U.K. S.E.R.C. for financial support.

References

1. Sawyer, L., Papiz, M. Z., North, A. C. T. & Eliopoulos, E. E. (1985) *Biochem. Soc. Trans.* **13**, 265–266
2. Newcomer, M. E., Jones, T. A., Åqvist, J., Sundelin, J., Eriksson, U., Rask, L. & Peterson, P. A. (1984) *EMBO J.* **3**, 1451–1454
3. Papiz, M. Z., Sawyer, L., Eliopoulos, E. E., North, A. C. T., Findlay, J. B. C., Sivaprasadarao, A., Jones, T. A., Newcomer, M. E. & Kraulis, P. (1986) *Nature* (*London*) **324**, 383–385
4. Pervaiz, S. & Brew, K. (1985) *Science* **228**, 335–337
5. Monaco, H. L., Zanotti, G., Spadon, P., Bolognesi, M., Sawyer, L. & Eliopoulos, E. E. (1987) *J. Mol. Biol.* **197**, 695–706
6. Huber, R., Schneider, M., Epp, O., Mayr, I., Messerschmitt, A., Pflugrath, J. & Kayser, H. (1987) *J. Mol. Biol.* **195**, 423–434
7. Huber, R., Schneider, M., Mayr, I., Muller, R., Deutzmann, R., Suter, F., Zuber, H., Falk, H. & Kayser, H. (1987) *J. Mol. Biol.* **198**, 499–513
8. Holden, H. M., Rypniewski, W. R., Law, J. H. & Rayment, I. (1987) *EMBO J.* **6**, 1565–1570
9. Jones, T. A., Bergfors, T., Sedzik, J. & Unge, T. (1988) *EMBO J.* **7**, 1597–1604
10. Sacchettini, J. C., Gordon, J. I. & Banaszak, L. J. (1988) *J. Biol. Chem.* **263**, 5815–5819
11. McRee, D. E., Tainer, J. A., Meyer, T. E., Van Beeumen, J., Cusanovich, M. A. & Getzoff, E. D. (1989) *Proc. Natl. Acad. Sci. U.S.A.* **86**, 6533–6537
12. Hendrickson, W. A., Pähler, A., Smith, J. L., Satow, Y., Merritt, E. A. & Phizackerley, R. P. (1989) *Proc. Natl. Acad. Sci. U.S.A.* **86**, 2190–2194
13. Godovac-Zimmermann, J. (1988) *Trends Biochem. Sci.* **13**, 64–66
14. Pevsner, J., Reed, R. R., Feinstein, P. G. & Snyder, S. H. (1988) *Science* **241**, 336–339
15. Sawyer, L. (1987) *Nature* (*London*) **327**, 659
16. Zagalsky, P. F. (1976) *Pure Appl. Chem.* **47**, 103–120
17. Zagalsky, P. F., Eliopoulos, E. E. & Findlay, J. B. C. (1990) *Comp. Biochem. Physiol.* **97B**, 1–18
18. Reference deleted
19. Reference deleted
20. North, A. C. T. (1989) *Int. J. Biol. Macromol.* **11**, 56–58
21. North, A. C. T. (1989) *J. Mol. Graphics* **7**, 67–70
22. Sivaprasadarao, A. & Findlay, J. B. C. (1988) *Biochem. J.* **255**, 561–569
23. Giguère, V., Ong, E. S., Evans, R. M. & Tabin, C. J. (1989) *Nature* (*London*) **337**, 566–569

Biochem. Soc. Symp. **57**, 49–64
Printed in Great Britain

Holistic Approaches to Receptor and Channel Structure and Dynamics

EDWARD M. KOSOWER*

Biophysical Organic Chemistry Unit, School of Chemistry, Sackler Faculty of Exact Sciences, Tel-Aviv University, Tel-Aviv 69978, Israel

Synopsis

A detailed three-dimensional structure of a molecule is a necessary, but not sufficient prerequisite to understanding its behaviour. Crystal structures are lacking for protein receptors and ion channels. However, the amino acid sequences for representatives of the most important superfamilies of neuroactive proteins (ligand-gated ion channels, voltage-gated ion channels and G-protein receptors) are known. To address the problem of three-dimensional structure and, subsequently, the dynamic properties of these molecules, a holistic approach was applied to the nicotinic acetylcholine receptor (nAChR), the sodium channel of nerves (NaC) and rhodopsin (Rh) to build 'working' structures. For nAChR and Rh, the chains are placed within an envelope derived from image analysis of electron micrographs. Physical organic principles suggest how to juxtapose binding groups as well as structures for ion channels, leading to the identification of the most common feature of channels, the positively charged 'S4' segment and the ever-present amphiphilic segment of nAChR proteins, which has homologous counterparts in the glutamate receptor and glutamate-binding proteins. The cyclic GMP-gated NaC (cGMP-NaC) is related to the eel and rat neuronal NaC, and has a novel positively charged control sequence.

Introduction

One prerequisite to understanding the behaviour of any molecule is a knowledge of its three-dimensional structure. Without the three-dimensional structure, it is quite difficult to make generally useful rules about structure and function. Crystals (and structures) are lacking for protein receptors and ion channels, owing to the difficulties common to the crystallization of large molecules. These are compounded by the presence of disparate domains in molecules having both hydrophobic and hydrophilic regions, notwithstanding the successful crystallization and crystal structure analysis of the bacterial reaction centre [1]. We can only hope that a few critical membrane proteins will be analysed with the required resolution [2], and can understand the major effort

* Also Department of Chemistry, State University of New York, Stony Brook, Stony Brook, NY 11794–3400, U.S.A.

put into bacteriorhodopsin using image analysis of electron scattering densities [3] and neutron diffraction [4,5].

The amino acid sequences for representatives of the most important superfamilies of neuroactive proteins (ligand-gated ion channels, voltage-gated ion channels and G-protein receptors) are known. If the linear sequences could be folded up by a suitable method, we might hope to obtain a 'working' two- or three-dimensional structure. To address the problem of three-dimensional structure and, subsequently, the dynamic properties of these molecules, we have used a holistic approach, meaning the use of any and all information to attack a problem. We have applied the technique to the nAChR [6], the NaC [7,8] and Rh [9], and have built 'working' structures. We will describe briefly each molecule and show how the analysis can be used for glutamate receptors and binding proteins and for the cGMP-NaC [10].

For nAChR and Rh, the chains are placed within an envelope derived from image analysis of electron micrographs. Physical organic principles suggest how to juxtapose binding groups as well as structures for ion channels, leading to the identification of the most common feature of channels, the positively charged 'S4' segment and the ever-present amphiphilic segment of nAChR proteins, that has homologous counterparts in the glutamate receptor and glutamate-binding proteins. The cGMP-gated sodium channel is related to the eel and rat neuronal NaC, and has a novel positively charged control sequence.

nAChR

nAChR consists of five polypeptide subunits ($\alpha_2\beta\gamma\delta$). The properties and principles used to arrange the sequences into a structure include (1) Hydrophobicity → membrane-crossing segments; (2) amphiphilic character → ion-carrying segments (ion channel with single group rotations) (Fig. 1); (3) molecular shape (elongated, pentagonal cylinder) → folding dimensions of exobilayer portion; (4) choice of acetylcholine-binding sites → specific folding of exobilayer segments; (5) location of reducible disulphides (near agonist-binding site) → additional specification of exobilayer arrangement; (6) genetic homology → consistency of functional group choices; (7) non-competitive antagonist labelling → arrangement of bilayer helices. The AChR model is divided into three parts: (*a*) exobilayer: 11 antiparallel β-strands from each subunit; (*b*) bilayer: 4 hydrophobic α-helices and 1 amphiphilic α-helix from each subunit; and (*c*) cytoplasmic: one (folded) loop from each subunit.

Lysine and glutamate (aspartate) side-chains are used to construct a model acetylcholine (ACh-) binding site in the acetylcholine receptor (AChR) to provide: (1) a counter charge for the positively charged trimethylammonio group and (2) a hydrogen bond to the acetyl carbonyl group. Lysine is the only reasonable choice for a full + (positive) charge at a well-defined position. At neutral pH, histidine is only partly protonated and the guanidino group of arginine has at least two (partial) + centres for binding to a nearby ligand. Conformationally mobile groups may be 'pinned' by interaction with other polar groups to lower the rate of spontaneous conformational change. Gluta-

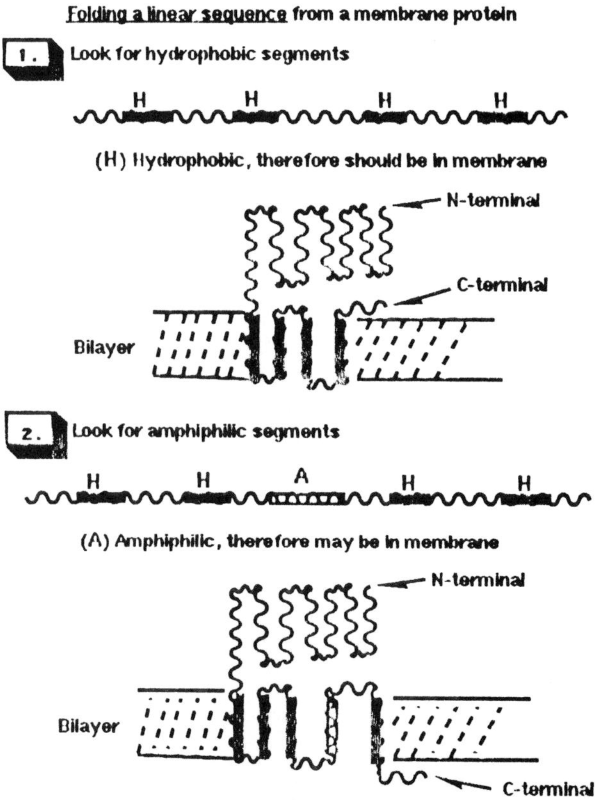

Fig. 1. *Schematic diagram illustrating two of the principles involved in the folding up of a linear sequence for a membrane protein*

The first principle involves the identification of hydrophobic segments. After the hydrophobic sequences are identified, they are inserted into the bilayer and must be separated by connecting segments. It is striking that many hydrophobic segments are about 24 amino acids long, corresponding to 36 Å of an α-helical segment and just the right size to bridge the usual phospholipid bilayer. The second principle requires the identification of amphiphilic segments. Such segments have polar groups along one side of either an α-helix or a 3_{10}-helix and largely hydrophobic groups along the other sides. The folding of a membrane protein located in the bilayer is shown.

mate [− (negative) charge] is used to 'pin' lysine (+ charge) in the present model. We assume that the binding site is between the two α-subunits and similarly arranged within each subunit. The binding site will thus consist of apposed lysine–glutamate (aspartate) sets.

The simplest conformational change which might have biological consequences was then sought after. The study of initial local motions, as small as single group rotations (SGR) with the receptor–ligand combination, turned out to be a useful approach to probing the behaviour of receptor sites, and for developing useful models for ion-carrying channels. The polar side-chain groups of proteins may be divided into three classes: (1) negatively charged, i.e. glutamate, aspartate; (2) positively charged, i.e. lysine, arginine and, possibly, histidine; (3) neutral, i.e. glutamine, asparagine and tyrosine. What types of motions might be expected for these groups? All of the side-chains named may

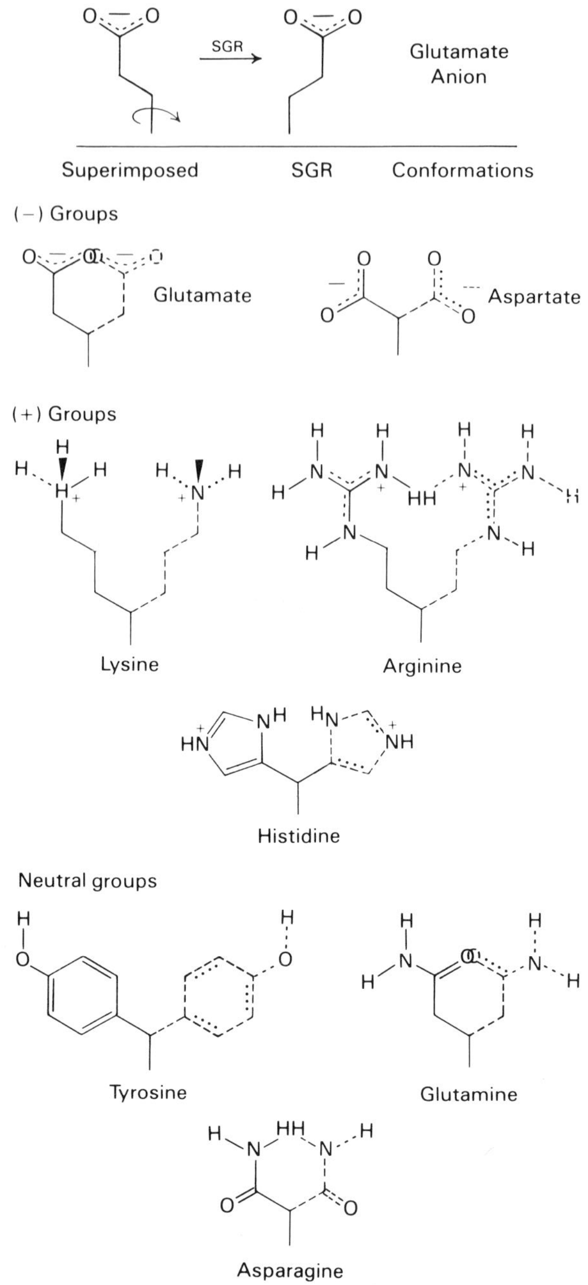

Fig. 2. *SGRs for polar amino acid side-chains of proteins*

A single group rotation for glutamate ion is given as an example of the small conformational change embodied in an SGR. Superimposed extreme forms of SGRs are shown for all of the ordinary polar amino acid side-chains found in proteins. The groups, according to charge, are (−) Glu, Asp, (+) Lys, Arg, His and (0) Tyr, Glu, Asn. Tyrosine is normally classified as a hydrophobic amino acid. Superimposition of the rotated form on top of the original conformation, using a dashed line for the group after single group rotation offers a clear view of the actual changes in the location of side-chain atoms (reprinted from [6], with permission).

undergo a number of local rotational changes. The maximum change represents a rotation of the whole group, which I shall call SGR. An SGR for the glutamate side-chain is shown at the top of Fig. 2, along with SGRs for other groups, using superimpositions of the initial and final conformations as compact representations of the SGR. Along an α-helix, an appropriate set of charged amino acids can interact in a way which can promote ion motion. This arrangement was postulated to occur in a channel and was then found in the sequence of the α-subunit of the nAChR. Since the segment contains a reasonable number of hydrophobic amino acids (Fig. 3), it was proposed as an amphiphilic channel element. A plausible model for a channel can be designed on this basis (Fig. 4).

The membrane portions of the nAChR account for most of the α-helices

BILAYER

398glu$^-$	427asp$^-$	398glu$^-$	439asn	445asn
397glu$^-$	426lys$^+$	397glu$^-$	438glu$^-$	444gly
396ala	425lys$^+$	396ala	437asn	443val
395ala	424leu	395ala	436glu$^-$	442glu$^-$
394asn	423asp$^-$	394asn	435ser	441glu$^-$
393ser	422asp$^-$	393ser	434gly	440asp$^-$
392ser	421phe	392ser	433ser	439tyr
391glu$^-$	420glu$^-$	391glu$^-$	432asp$^-$	438ala
390glu$^-$	419ser	390glu$^-$	431asn	437asn
389asp$^-$	418ala	389asp$^-$	430gln	436lys$^+$
388ser	417ser	388ser	429glu$^-$	435glu$^-$
387lys$^+$	416glu$^-$	387lys$^+$	428lys$^+$	434lys$^+$
386met	415leu	386met	427thr	433ile
385his	414gln	385his	426ser	432gln
384glu$^-$	413glu$^-$	384glu$^-$	425lys$^+$	431lys$^+$
383ala	412ala	383ala	424ala	430val
382ile	411ile	382ile	423ile	429ile
381tyr	410tyr	381tyr	422phe	428tyr
380lys$^+$	409lys$^+$	380lys$^+$	421asn	427asn
379val	408ile	379val	420cys	426thr
378gly	407ala	378gly	419ala	425ser
377glu$^-$	406glu$^-$	377glu$^-$	418glu$^-$	424asp$^-$
376ile	405val	376ile	417val	423ile
375ala	404ala	375ala	416cys	422gly
374ser	403glu$^-$	374ser	415ser	421ser
373lys$^+$	402lys$^+$	373lys$^+$	414lys$^+$	420lys$^+$
372val	401leu	372val	413ile	419ile
371asp$^-$	400asp$^-$	371asp$^-$	412glu$^-$	418glu$^-$
370pro	399gln	370pro	411pro	417asp$^-$
369asn	398pro	369asn	410ala	416his
α	β	α	γ	δ

Fig. 3. *Ion channel elements, one from each subunit, of the nAChR*

The elements are aligned in the way needed for an active channel like that in Fig. 4 (reprinted from [6] with permission).

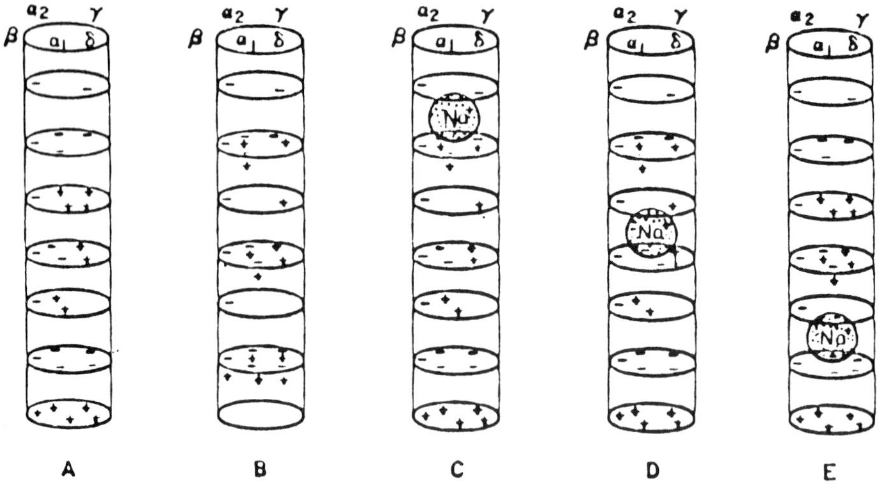

Fig. 4. *A schematic model for the nAChR ion channel*

(A) Charge distribution based on Fig. 3. (B) Charge distribution after allowing for Coulombic attaction and SGR. For simplicity, only + charges have been moved. (C) Entry of a hydrated Na^+ ion into the channel. (D) Further movement of the Na^+ ion. (E) A particularly favourable location for the Na^+ ion. Not all stages of the Na^+ ion motion are illustrated (reprinted from [6] with permission).

within the protein. About half of the receptor is exobilayer, and can be modelled on the basis of the molecular shape [11–15], the assignment of the substantial β-strand content to this region, and the choice of the binding sites as sets of +, − groups, with one near the reducible disulphide. A scheme for the exobilayer portion of the α-subunit is shown in Fig. 5, and an overall model for the α-subunit is shown in Fig. 6.

One of the most characteristic properties of the nAChR receptor, desensitization, can be explained by a change in the orientation of the ACh ligand or perhaps by a change to a more compact conformation, as indicated by n.m.r. studies [16]. The 'flower' model explains the activation of the receptor (Fig. 7), admitting ions to the channel, and closing the channel with desensitization or absence of the ACh ligand (Fig. 8). The details of the holistic approach to the structure of the nACHR are summarized in Table 1 and the overall approach to nAChR, the NaC and Rh are given in Table 2.

The functional expression technique used by Masu *et al.* [17] for the substance K receptor has been utilized to isolate the glutamate receptor [18]. Frog and chick kainate-binding proteins have been isolated, peptides produced, suitable probes for cDNA generated and the sequences obtained [19,20]. All three proteins appear to contain four transmembrane segments and some other features which suggest that they belong to the superfamily of ligand-gated receptor proteins. The receptor is functional in *Xenopus* öocytes [18]; tests of one binding protein did not show ion channel activity [19]. To strengthen the case for similarity among the three glutamate proteins, an

```
  157ser 156val 155lys+
 158ile          154thr                                             82ser     81pro
159ser            153gly                                            83asp-    80leu     41ile    40leu    1ser
160pro            152asp-                                           84asp-    79arg+    42ser    39gln    2glu-
161glu-            151tyr                     119thr                 85val     78ile     43val    38ile    3his
162ser     189tyr--150thr--190tyr    120pro           118trp        86trp     77lys+    44asp-   37leu    4glu-
163asp-   188val    149trp    191thr    121pro        117met        87leu     76lys+    45glu-   36gln    5thr
164arg+   187trp    148ile    192cys|   122ala        116ile        88pro     75ile     46val    35leu    6arg+
165pro    186his    147gly    193cys|   123ile        115lys+       89asp-    74gly     47asn    34gly    7leu
166asp-   185lys+   146leu    194pro    124phe        114gly        90leu     73gly     48gln    33val    8val
167leu    184trp    145lys+   195asp-   125lys+       113thr        91val     72tyr     49ile    32thr    9ala
168ser    183gly    144met    196thr    126ser        112tyr        92leu     71asp-    50val    31ile    10asn
169thr    182arg+   143thr    197pro    127tyr        111asp-       93tyr     70ala     51glu-   30asp-   11leu
170phe    181tyr    142cys---198tyr--128cys           110leu        94asn     69pro     52thr    29val    12leu
171met    180asp-   141asn    199leu    129glu-       109leu        95asn     68asn     53asn    28phe    13glu-
172glu-   179lys+   140gln    200asp-   130ile        108leu        96ala     67trp     54val    27his    14asn
173ser    178met    139gln    201ile    131ile        107lys+       97asp-    66arg+    55arg+   26thr    15tyr
174gly     177val    138asp-   202thr    132val       106thr        98gly     65leu     56leu    25his    16asn
175glu-    176trp    137phe    203tyr    133thr       105met        99asp-    64arg+    57arg+   24his    17lys+
                    136pro    204his    134his        104his        100phe    63val     58gln    23glu-   18val
                      135phe-205phe --|               103val        101ala    62asp-    59gln    22val    19ile
                              206ile                      102ile              61ile     60trp    21pro    20arg+
                              207met
                              208gln
                              209arg+

 X9       X10       X8        X11       X7            X6            X5        X4        X3       X2       X1
```

Fig. 5. *The exobilayer arrangement of the nAChR α-subunit according to the criteria explained in the text*

Reprinted from [6], with permission.

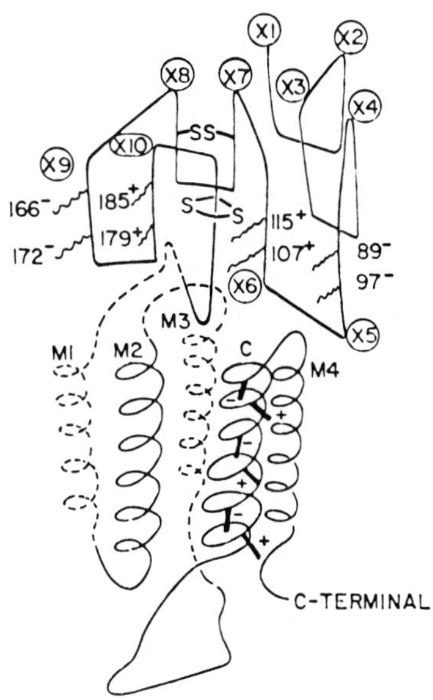

Fig. 6. *Relationship of exobilayer strands (X1–X10, X11 with Cys–Cys–SS bond not marked) to the bilayer helices (M1–M3, C = channel, M4) and the cytoplasmic segments in an α-subunit*

Binding site groups are named, and the charges on the ion channel element indicated (reprinted from [6], with permission).

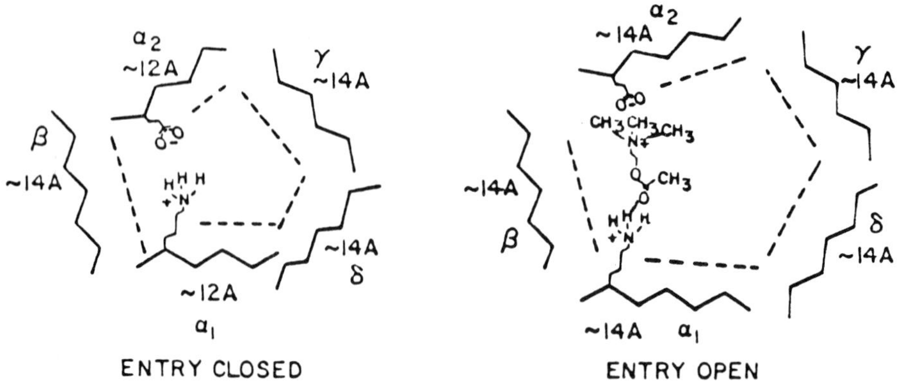

Fig. 7. *Schematic drawings of the resting (closed) and active (open) forms of the nAChR*

The α-exobilayer strands approach the centre of the 'cup' more than the other strands. The charges on the αβα side of the pentagon are more balanced than those on the γδ side (reprinted from [6], with permission).

attempt was made to establish additional homologies on the basis of the nAChR amphiphilic element. By starting with the α-amphiphilic sequence (the α-subunit ion channel element), homologies were sought between the nAChR and the glutamate proteins, using the idea that the amphiphilic segment, which has been proposed as a channel element, was present. The homologies were indeed present (Fig. 9). Along the putative channel element, the binding protein differs from the receptor in having proximate lysines. A plausible interpretation of the difference between the receptor and binding proteins is that the latter require larger potential changes to diminish the interaction between

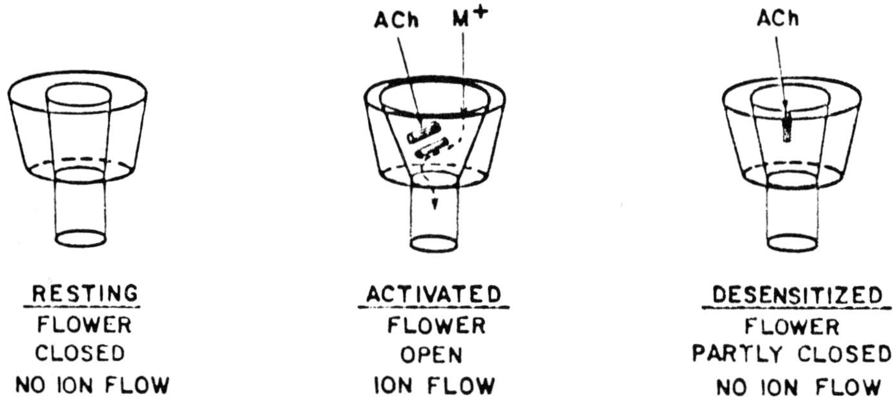

Fig. 8. *Three-dimensional model ('flower'*) representing in the simplest possible way the operation of the nAChR*

The resting state has an inner set of strands (inner portion of the cup) close enough so as to block ion influx. Entry of two ACh molecules in an arrangement perpendicular to the strands forces open the inner cup, creating the activated state and revealing the ion pathway. After sometime, the agonist (ACh) molecules rearrange to a parallel (or bent) position, the cup partially closes, forming the desensitized state, with the ion pathway blocked. * Usually a blossom, but can be read as something which flows. Reprinted from [6], with permission.

Table 1. *Holistic analysis of the nAChR*

Physical and chemical data* ⋙			Exobilayer portion		→	Exobilayer model
Amino acid sequence	Functional model**	→	Ion channel elements	AChR bilayer helices	→	Bilayer model
	Hydrophobicity measure	→	Hydrophobic helices			
Physical and chemical data*** ⋙			Cytoplasmic portion		→	Cytoplasmic model

* Electron microscopy, X-ray scattering, functional model and SGR theory, ACh-binding sites, thiol labelling, Raman data, genetic homology, deletion and mutation data.

** Ion channel mechanism and structure based on SGR theory, genetic homology, deletion and mutation data, non-competitive antagonist labelling data.

*** Electron microscopy, immunological data, phosphorylation sites.

Table 2. *Holistic analysis of protein structure*

EM, electron microscopy.

Protein	Structural envelope	Functional groups	Membrane elements
nAChR	Image analysis (EM)	+, − Binding channel model	Hydrophobicity Amphiphilic segments
Rh	Bacteriorhodopsin image analysis (EM)	Lys, point charge model	Hydrophobicity Location of linking Lys
NaC	No reference structure	Amphiphlic 3_{10}-helix	Hydrophobicity

helix-adjacent lysines or, as suggested elsewhere [18,19], other subunits are required for activity.

NaC

The revelation of the eel NaC sequence stimulated a comparison with the nAChR. Alternating charge-carrying segments were not found, but rather an amphiphilic segment recognized by Numa and co-workers as a potential

R 628
P - **I** V <u>**S**</u> A **E** **D** **L** A **K** **Q** T E I A Y **G** **T** **L** E A G **S** **T** K E **F** **F** R R **S** **K** I 660

B1 261
P N **I** Q <u>**S**</u> V **E** **D** **L** L **K** **Q** D K L D F **G** **T** **L** S N S **S** **T** L N **F** **F** K N **S** **K** N P 295

B2 244
L S **I** Q <u>**T**</u> F **E** **D** **L** V **K** **Q** R K L E F **G** **T** **L** D G S **S** **T** F Y **F** **F** K N **S** **K** N P 278

nAChR 370α
P D <u>**V**</u> K <u>**S**</u> A I <u>**E**</u> G <u>**V**</u> **K** Y I A E H M K <u>**S**</u> D <u>**E**</u> E S **S** N A A E E 398α

Fig. 9. *A comparison of sequences from the glutamate receptor and the two glutamate receptor-binding proteins with the amphiphilic sequence from the α-subunit of the nAChR*

The homologies among the first three are quite clear from the amino acids shown in bold face and to the nAChR by inspection. The underlined amino acids represent functionally plausible substitutions (E for D, etc.).

210 arg$^+$	657 arg$^+$	1093 ala	1417 arg$^+$	910 asp$^-$	942 glu$^-$
211 thr	658 ser	1094 ile	1418 val	909 glu$^-$	943 glu$^-$
212 phe	659 leu	1095 lys$^+$	1419 ile	908 glu$^-$	944 glu$^-$
213 arg$^+$	660 arg$^+$	1096 asn	1420 arg$^+$	907 asp$^-$	945 glu$^-$
214 val	661 leu	1097 leu	1421 leu	906 ser	946 glu$^-$
215 leu	662 leu	1098 arg$^+$	1422 ala	905 ser	947 glu$^-$
216 arg$^+$	663 arg$^+$	1099 thr	1423 arg$^+$	904 asp$^-$	948 glu$^-$
217 ala	664 ile	1100 ile	1424 ile	903 val	949 glu$^-$
218 leu	665 phe	1101 arg$^+$	1425 ala	902 leu	950 glu$^-$
219 lys$^+$	666 lys$^+$	1102 ala	1426 arg$^+$	901 gly	951 pro
220 thr	667 leu	1103 leu	1427 val	900 glu$^-$	952 glu$^-$
221 ile	668 ala	1104 arg$^+$	1428 leu	899 glu$^-$	953 glu$^-$
222 thr	669 lys$^+$	1105 pro	1429 arg$^+$	898 glu$^-$	954 leu
223 ile	670 ser	1106 leu	1430 leu	897 ile	955 glu$^-$
224 phe	671 trp	1107 arg$^+$	1431 ile	896 glu$^-$	956 ser
225 pro	672 pro	1108 ala	1432 arg$^+$	895 ser	957 lys$^+$
226 gly	673 thr	1109 leu	1433 ala	894 glu$^-$	958 asp$^-$
227 leu	674 leu	1110 ser	1434 ala	893 gly	959 pro
$^1C^+$	$^2C^+$	$^5C^+$	$^6C^+$	$^3C^-$	$^4C^-$

Fig. 10. *Eel electroplax NaC elements*

Reprinted from [7], with permission.

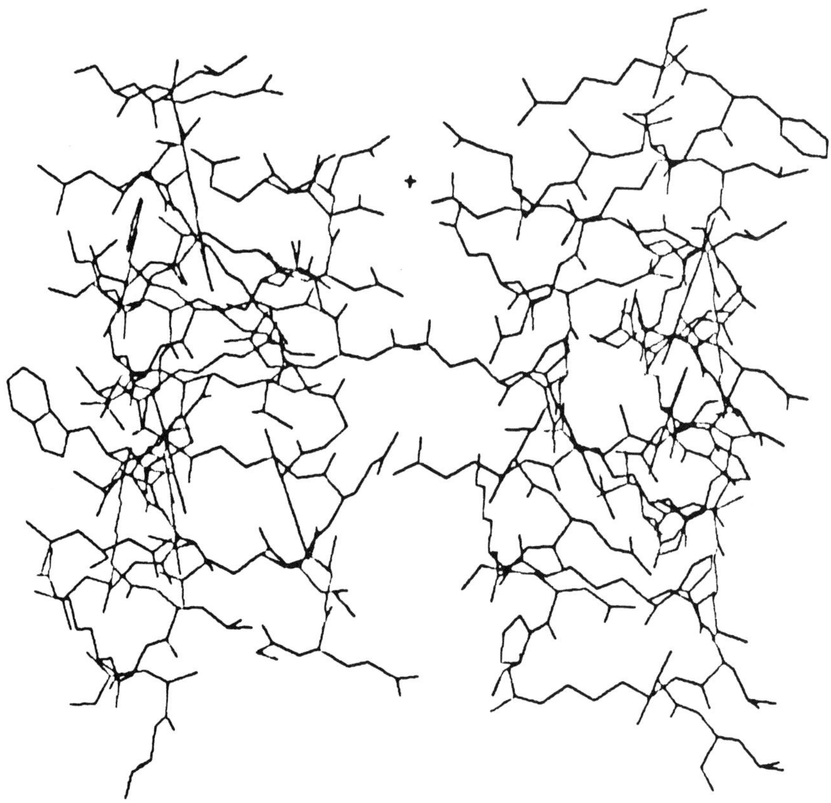

Fig. 11. *A side view of a 'stick' model of the NaC, looking at the central part of the NaC*

The (+) sign at the top between the two central helices represents the Na^+ ion with the sodium-binding site. Almost every side-chain on the two central helices carries a carboxylic acid group (glutamate), while most of the chains which project towards the centre are guanidino (arginine). Reprinted from [7], with permission.

3_{10}-helix [21,22]. We assigned the 3_{10}-helix segments as channel elements (Fig. 10). Meiri *et al.* [23] have shown that antibodies to the first peptide segment of this type in the sequence (C_1+) alter the electrophysiology of fast NaC in dorsal root ganglion cells. A negatively charged long segment was used to provide local charge compensation (Fig. 11). We could show that the model channel is blocked by cocaine (schematic for lidocaine blockage in Fig. 12) at what we think is the local anaesthetic-binding site [7].

By looking in the region of the functional segments just described, sufficient homology between the usual NaC and the cGMP-NaC was found. The sequences are shown in Fig. 13, with homologies starred. A remarkable segment of charged amino acids occurs in the *N*-terminal region. If these are *β*-strands, one finds the most surprising distribution for the charges; on one side, there are mostly ion pairs, and on the other, mostly positive charges (Fig. 14). Combination with cyclic GMP or phosphorylation would neutralize the charge and permit passage of positive charges. Thus, while cyclic GMP is available, the NaC could function in the usual way. In the absence of cyclic GMP (hydrolysed by the enzyme activated by light-activated Rh via a G-protein segment), the NaC would not be activated and the absence of a signal

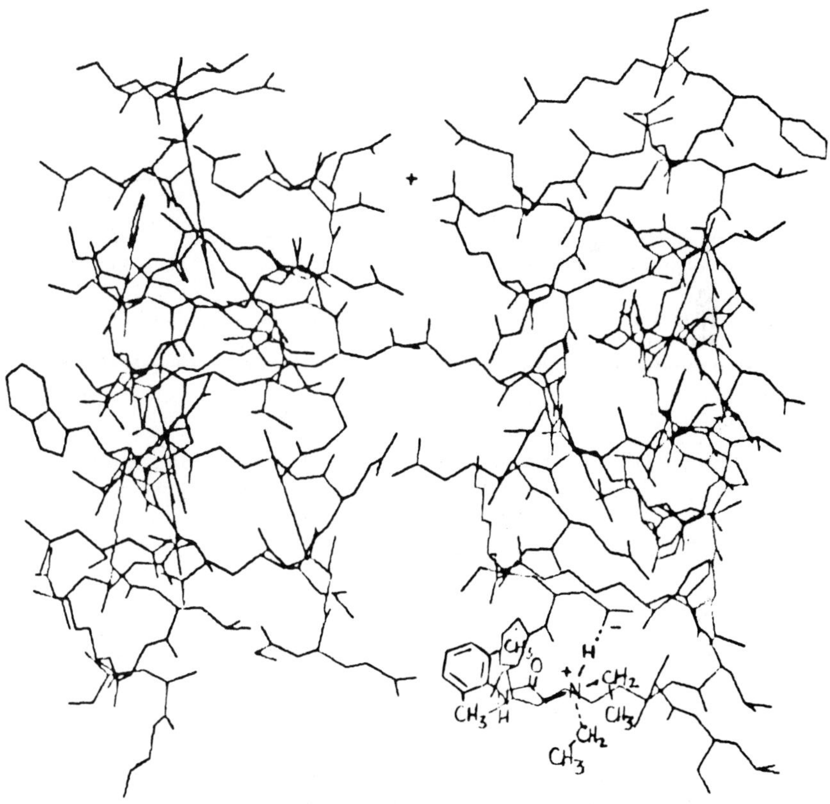

Fig. 12. *Side view of a 'stick model' of the eel NaC along with a molecule of lidocaine, a local anaesthetic*

Reprinted from [7], with permission.

Comparison of sodium channel sequences to bcG-NAC segments

Sodium channel "functional" sequences:
eel, rat brain I, II, bovine cGMP-gated

3_{10}-helices				negative strands
			1418 VVVI	*912 NEED 577
*209 LLLI 268	*653 MLLI	*1094 IIII	*1419 IIII	*913 KEED 578
*210 RRRR 269	654 SSSR	*1095 KKKR	*1420 RRRR	914 KSSL 579
211 TTTL 270	*655 VVVL	1096 NSSL	*1421 LLLL	915 KKKM 580
212 FFFN 271	656 LLLN	1097 LLLN	1422 AAAN	*916 HEEE 581
*213 RRRR 272	*657 RRRR	*1098 RRRR	*1423 RRRR	*917 AKKA 582
*214 VVVL 273	658 SSSL	1099 TTTL	*1424 IIIL	*918 LLLL 583
*215 LLLL 274	*659 LFFL	*1100 ILLL	1425 AGGL	919 NNNT 584
*216 RRRR 275	*660 RRRR	*1101 RRRR	*1426 RRRR	
*217 AAAI 276	*661 LLLI	*1102 AAAI	*1427 VIII	*945 EEEE 585
218 LLLS 277	662 LLLS	1103 LLLS	1428 LLLS	946 EQQY 586
*219 KKKR 278	*663 RRRR	*1104 RRRR	*1429 RRRR	*947 EPPP 587
220 TTTM 279	*664 IVVM	1105 PPPM	*1430 LLLM	*948 EVED 588
*221 IIIF 280	*665 FFFF	*1106 LLLF	*1431 IIIF	*949 EMAA 589
222 TSSE 281	666 KKKE	1107 RRRE	1432 RKKE	950 EEEK 590
*223 IVVF 282	*667 LLLF	*1108 AAAF	1433 AGGF	951 PPPG 591
*224 FIIF 283	*668 AAAF	*1109 LLLF	*1434 AAAF	952 EEEM 592
225 PPPQ 284	669 KKKQ	*1110 SSSQ	1435 KKKQ	953 EEE- 593
226 GGGR 285	670 SSSR	*1111 RRRR	1436 GGGR	*954 L--L 594
227 LLLT 286	671 WWWT	1112 FFFT	1437 IIIT	*955 E--E 595
228 KKKE 287	672 PPPE	*1113 EEEE	1438 RRRE	956 STSE 596
*229 TTTT 288	*673 TTTT	1114 GGGT	*1439 TTTT	*957 KLLK 597
				958 DEEG 598
				959 PPPK 599
				960 EEEQ 600

Fig. 13. *The amphiphilic segments assigned as 3_{10}-helices in the eel, rat I and rat II NaC are compared with an apparently similar sequence from the cGMP-NaC from the rod cell of the bovine retina*

For ease of comparison, the cGMP-NaC is given next to each triplet from the other NaC. A sequence with many negatively charged amino acids located after H6 is compared with two somewhat separated sequences from the centre of the other NaC proteins. Eel channel numbering is used for the NaC channels.

would be interpreted by the following cells as a sign for the absorption of a photon. A scheme for the functional segments of the cGMP-NaC is given in Fig. 15.

Rh

Rh, the light-receptor protein, is assigned a structure based on seven helical transmembrane segments, selected by analysis of the amino acid sequence [24–26]. There is thus a relationship to bacteriorhodopsin, a protein with an identical chromophore and for which an elliptical arrangement of seven transmembrane helices has been established by image analysis of electron micrographs and neutron diffraction. The chromophores are in the middle of the bilayer. Rh undergoes in stages a number of conformational changes to metarhodopsin II which then initiates a cascade of reactions through inter-

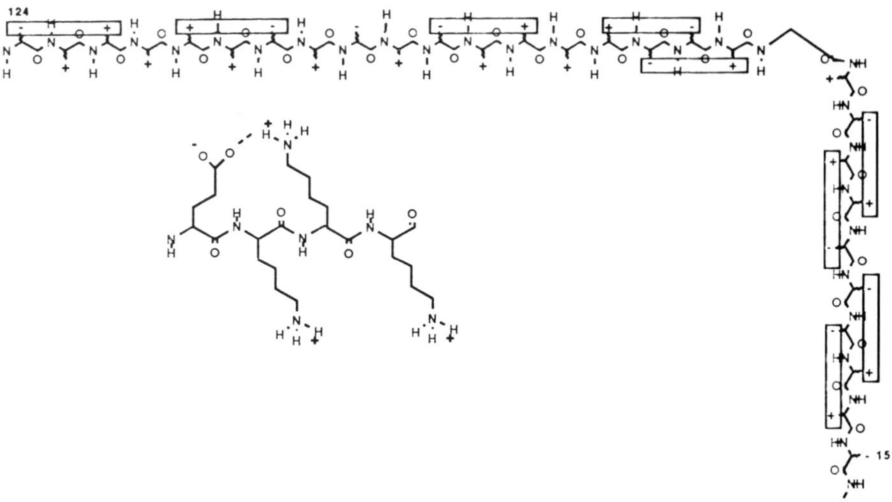

Fig. 14. *The unusual segments from the N-terminal portion of the cyclic cGMP-NaC from the rod cell of the bovine retina are illustrated in the form of β-strands*

Remarkably, on one side, most of the charges can form ion pairs, as shown in the central formula; on the other side mostly positive charges occur. These seem suitable as 'control elements', retarding the flow of cations like Na^+ if not neutralized through the intervention of negatively charged cyclic GMP or a phosphorylated protein. Photons absorbed by Rh eventually cause the hydrolysis of cyclic GMP and the failure of the Na^+ entry, thus constituting a negative signal to the neuron following the rod cell.

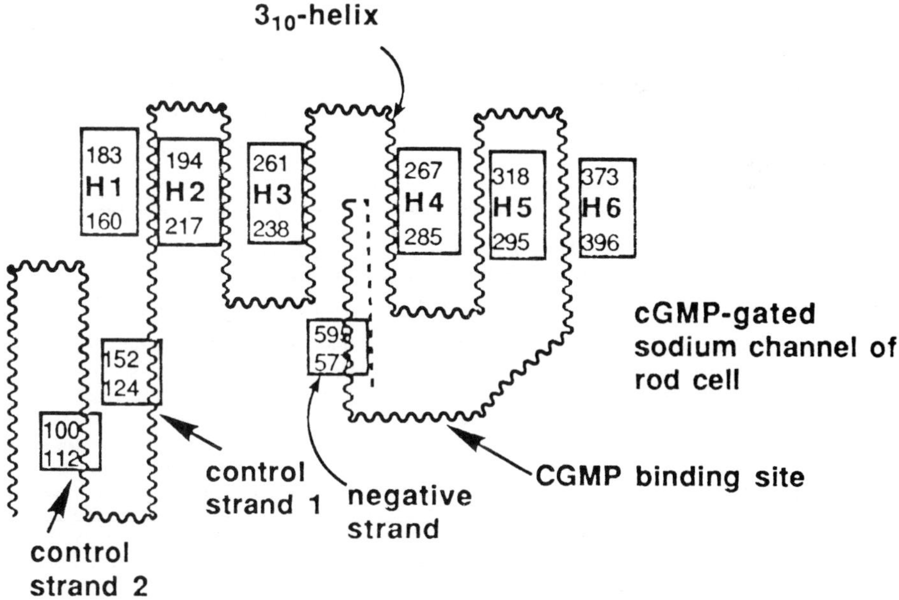

Fig. 15. *A schematic representation of the structure of the cGMP-NaC with the various portions labelled*

Strikingly, the sequence that carries the negative charge that may neutralize the positive charges of the 3_{10}-helices in the bilayer is located next to the site assigned as a cyclic GMP-binding site on the basis of homology to cyclic GMP binding proteins.

action with a G-protein. The absorption maximum of Rh is located at wavelengths favourable for daylight sensing, and at a position longer than might have been expected for the structure of the chromophore, the 11-*cis*-retinylidene iminium ion. The difference has been termed the 'opsin shift' and has been explained by the 'point-charge' model [27]. In the model, strengthened

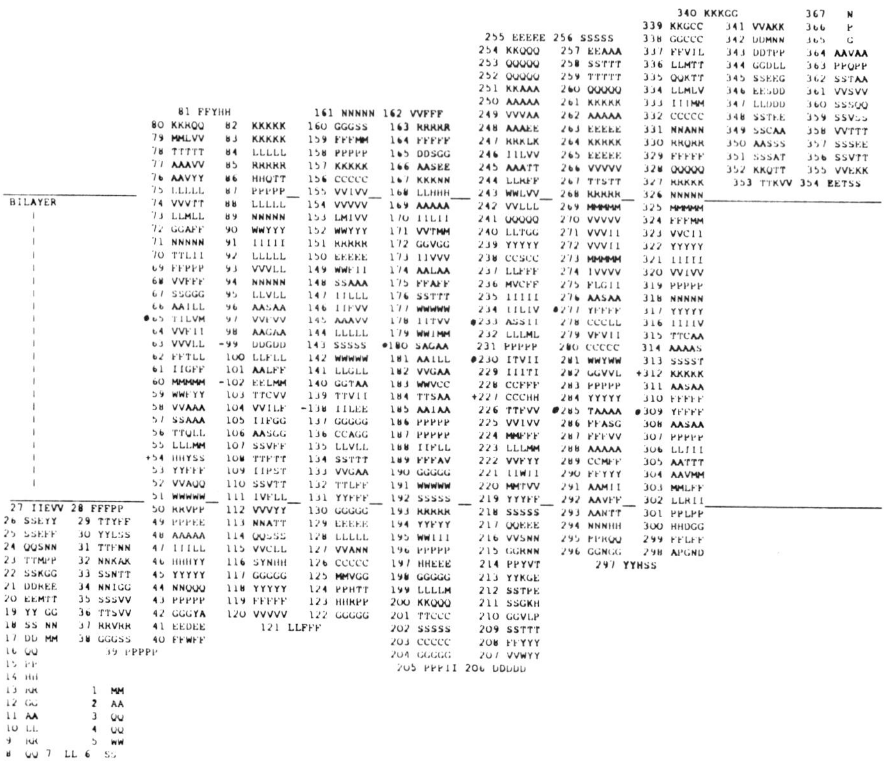

Fig. 16. *Amino acid sequences and schematic structures of human red, green, and blue iodopsins (cone pigments) and human and bovine rhodopsins (rod pigments) constructed using the single letter code for amino acids*

The distribution of the protein between the membrane bilayer and the aqueous interior of the rod disk (cone invagination) is indicated. The *N*-terminal residue of the opsin is in the interior of the disk (corresponding to the cell exterior of a cone), at the lower part of the Figure. Aligned sequences are shown using the homology numbering based on red–green iodopsins. To obtain the position in the Rh sequence, subtract 16 (19 for numbering of the blue iodopsin). No skips are required for the alignment. Membrane bilayer segments of 24 amino acids (36 Å) are chosen on the basis of regions of hydrophobicity in the aligned sequences. Seven segments are found, a result that agrees with more formal methods of choosing hydrophobic segments. It is assumed that these segments occur as α-helices (numbered 1–7). The 312 Lys bonded to the retinylidene moiety as an iminium ion, is located in the middle of helix 7 and is common to all known opsins. The second charged group common to four of the five opsins is Asp-99, which must be the counter-ion to the iminium ion. The next most salient feature is the Glu-138 found in the rhodopsins. At the same level in both rhodopsins is His-227, which thus is a good choice as a counter-ion (in protonated form) to Glu-138. In red–green iodopsins, Glu-102 is found at a location appropriate for pairing (weakly) with His-54. No counter-ion is found for the iminium ion in blue iodopsin. Amino acids with OH groups with an influence on the spectroscopic properties of the rhodopsins are marked with a black dot (reprinted from [9], with permission).

by homologies with cone pigments, the iodopsins [28], several charges and polar groups appear to be located in the middle of the bilayer near the chromophore [9]. The schematic model is shown in Fig. 16. Criticism of the point-charge model has been based on the failure of mutagenesis of some of the groups to have the expected effect on the opsin shift [29,30]. Two counter arguments are: (1) from the structural point of view, we would presume that the mid-bilayer polar 'belt' has a function, a point emphasized by the change to a 470 nm maximum for a 4-hydroxyretinylidene opsin [31]; (2) one of the reported experiments (with E113Q) [29] [E129 with the iodopsin number used in Fig. 16] implies that the pK_a of the rhodopsin imine has shifted [32] from above 10 to about 6.0, a change that implies that natural Rh has a specially stabilized ground state. This would cause the iminium ion absorption to occur at shorter wavelengths, not longer, than the mutant, since it would take more energy to excite the molecule. Deprotonation of the imine occurs at pH 7.2 in the mutant. E113 may thus serve another function, perhaps ion-pairing with R177 to create a tight binding pocket for the retinylidene group with minimum exposure to water. Retinylidene protonation is quite sensitive to the amount of water present in non-polar environments [33]. That apparently innocuous substitutions can have substantial effects on binding energies, and consequently on conformation, has been shown in the comparison of the binding of hen egg-white lysozyme (Q121) and California quail egg-white lysozyme (H121) with the antibody for the former. Molecular graphics studies of the crystal structures had suggested that both combinations would have equivalent stabilities [34].

Greg Petsko, Brandeis University and Richard J. Feldmann, N.I.H. are thanked for their patient help with some of the basic molecular graphics.

References

1. Deisenhofer, J., Epp, O., Miki, K., Huber, R. & Michl, H. (1985) *Nature (London)* **318**, 618–624
2. Eisenberg, D. & Hill, C. P. (1989) *Trends Biochem. Sci.* **14**, 260–264
3. Henderson, R. (1977) *Annu. Rev. Biophys. Bioeng.* **6**, 87–109
4. Popot, J.-L., Engelman, D. M., Gurel, O. & Zaccaï, G. (1989) *J. Mol. Biol.* **210**, 829–848
5. Dencher, N. A., Dresselhaus, D., Zaccai, G. & Büldt, G. (1989) *Proc. Natl. Acad. Sci. U.S.A.* **84**, 5058–5062
6. Kosower, E. M. (1987) *Eur. J. Biochem.* **168**, 431–449
7. Kosower, E. M. (1991) *Molecular Mechanisms for Sensory Signals: Recognition and Transformation*, Princeton University Press, Princeton, New Jersey, U.S.A. in the press
8. Kosower, E. M. (1985) *FEBS Lett.* **182**, 234–242
9. Kosower, E. M. (1988) *Proc. Natl. Acad. Sci. U.S.A.* **85**, 1076–1080
10. Kaupp, U. B., Nildome, T., Tanabe, T., Terada, S., Bönigk, W., Stühmer, W., Cook, N. J., Kangawa, K., Matsuo, H., Hirose, T., Miyata, T. & Numa, S. (1989) *Nature (London)* **342**, 762–766
11. Kistler, J., Stroud, R. M., Klymkowsky, M. W., LaLancette, R. & Fairclough, R. H. (1982) *Biophys. J.* **37**, 371–383
12. Brisson, A. & Unwin, P. N. T. (1985) *Nature (London)* **315**, 474–477
13. Toyoshima, C. & Unwin, N. (1988) *Nature (London)* **336**, 247–250
14. Unwin, N., Toyoshima, C. & Kubalek, E. (1988) *J. Cell Biol.* **107**, 1123–1138
15. Giersig, M., Kunath, W., Sack-Kongehl, H. & Hucho, F. (1986) in *Nicotinic Acetylcholine Receptor: Structure and Function* (Maelicke, A., ed.) vol. H-3, pp. 7–17, NATO ASI Series, 1986, Springer, Berlin

16. Behling, R. W., Yamane, T., Navon, G. & Jelinski, L. W. (1988) *Proc. Natl. Acad. Sci. U.S.A.* **85**, 6721–6725
17. Masu, Y., Nakayama, K., Tamaki, H., Harada, Y., Kuno, M. & Nakanishi, S. (1987) *Nature* (*London*) **329**, 836–838
18. Hollmann, M., O'Shea-Greenfield, A., Rogers, S. W. & Heinemann, S. (1989) *Nature* (*London*) **342**, 643–648
19. Wada, K., Dechesne, C. J., Shimasaki, S., King, R. G., Kusano, K., Buonanno, A., Hampson, D. R., Banner, C., Wenthold, R. J. & Nakatani, Y. (1989) *Nature* (*London*) **342**, 684–689
20. Gregor, P., Mano, I., Maoz, I., McKeown, M. & Teichberg, V. I. (1989) *Nature* (*London*) **342**, 689–692
21. Noda, M., Shimizu, S., Tanabe, T., Takai, T., Kayano, T., Ikeda, T., Takahashi, H., Nakayama, H., Kanaoka, Y., Minamino, N., Kangawa, K., Matsuo, H., Raftery, M. A., Hirose, T., Notake, M., Inayama, S., Hayashida, H., Miyata, T. & Numa, S. (1984) *Nature* (*London*) **312**, 121–127
22. Noda, M., Ikeda, T., Kayano, T., Suzuki, H., Takeshima, H., Kurasaki, M., Takahashi, H. & Numa, S. (1986) *Nature* (*London*) **320**, 188–192
23. Meiri, H., Spira, G., Marei, S., Namir, M., Schwartz, A., Komoriya, A., Kosower, E. M. & Palti, Y. (1987) *Proc. Natl. Acad. Sci. U.S.A.* **84**, 5058–5062
24. Hargrave, P. A., McDowell, J. H., Feldmann, R. J., Atkinson, P. H., Mohana Rao, J. K. & Argos, P. (1984) *Vision Res.* **24**, 1487–1499
25. Abdulaev, N. G., Artamonov, I. D., Bogachuk, A. S., Feigina, M. Yu., Kostina, B., Kudelin, A. B., Martynov, V. I., Miroshnikov, A. I., Zolotarev, A. S. & Ovchinnikov, Yu. A. (1982) *Biochemistry Int.* **5**, 693–703
26. Dratz, E. A. & Hargrave, P. A. (1983) *Trends Biochem. Sci.* **8**, 128–131
27. Honig, B., Dinur, B. & Nakaniski, K., Balogh-Nair, V., Gawinowicz, M. A., Arnaboldi, M. & Motto, M. G. (1979) *J. Am. Chem. Soc.* **101**, 7084–7086
28. Nathans, J., Thomas, D. & Hogness, D. S. (1986) *Science* **232**, 193–202
29. Zhukovsky, E. A. & Oprian, D. D. (1989) *Science* **246**, 928–930
30. Nathans, J. (1990) *Biochemistry* **29**, 937–942
31. Corson, D. W., Cornwall, M. C., MacNichol, E. F., Mani, V. & Crouch, R. K. (1990) *Biophys. J.* **57**, 109–115
32. Albeck, A., Friedman, N., Ottolenghi, M. & Sheves, M. (1986) *Proc. Natl. Acad. Sci. U.S.A.* **83**, 3262–3266
33. Singh, A. K., Sandorfy, C. & Fendler, J. H. (1990) *J. Chem. Soc. Commun.* 233–234
34. North, A. C. T. (1989) *J. Mol. Graphics* **7**, 67–70

Biochem. Soc. Symp. **57**, 65–79
Printed in Great Britain

Modelling of Binding Sites of the Nicotinic Acetylcholine Receptor and their Relation to Models of the Whole Receptor

VICTOR B. COCKCROFT*†, GEORGE G. LUNT† and DAVID J. OSGUTHORPE*

* *Molecular Graphics Unit, University of Bath, Claverton Down, Bath BA2 7AY, U.K. and*
† *Biochemistry Department, University of Bath, Claverton Down, Bath BA2 7AY, U.K.*

Synopsis

Models for the acetylcholine (ACh)-binding site of the nicotinic acetylcholine receptor (nAChR) are proposed. These models have been developed by using the concept of the ligand-gated ion-channel (LGIC) superfamily of receptors that have evolved from a common ancestor. An initial component of the binding site was identified as a highly conserved 15-residue stretch of primary structure in the *N*-terminal extracellular region of all known LGIC subunits, based on aligned sequence data of LGICs. This subregion, termed the Cys-loop, was modelled as an amphiphilic β-hairpin and we propose that it forms a major determinant of the binding cleft for agonists. This initial, partial binding-site model has been extended to include residues biochemically identified as spatially adjacent to the binding cleft.

A recently developed technique for rapidly scanning the known protein structural database for 'non-homologous similarity' using just sequence information identified the known structure of the enzyme pyrophosphatase (PPase) as a candidate scaffold for the *N*-terminal domain of the nAChR. This similarity was investigated further using sequence alignments. A framework model of the full *N*-terminal domain in which the position of the Cys-loop and other binding-site determinants, as well as the main immunogenic region (MIR), have been mapped on to the PPase structure.

Introduction

The nAChR is the most well-studied member of the LGIC superfamily of proteins. As such, it is the prototype for understanding conserved structure–function relationships within the extended LGIC class of receptors. In response to the binding of agonist, the nAChR permits a flux of cations across the membrane through an ion-channel integral to its structure. The superfamily presently encompasses, in addition to the nAChR [1], the receptors for γ-aminobutyric acid (of the $GABA_A$ type) [2] and Gly [3]. The latter two receptors function in a similar way, but the channels are selective for anions. The nAChR is composed of five homologous subunits arranged symmetrically to give a barrel-shape with a cylindrical extracellular funnel and a centrally located ion-channel [4]. However, currently, the highest-resolution information available for this receptor is at approximately 8–15 Å resolution [5,6], which only gives an idea of the outline shape of the receptor.

The lack of definitive structural information for these receptors and, indeed, any definitive information about the location of the binding site makes any form of ligand design extremely difficult. Currently, although X-ray crystallography has shown it is now relatively easy to determine the structure of a globular protein, given that crystals of the protein can be grown, only a few structures of transmembrane proteins have been solved. The only high-resolution structure for a transmembrane protein that has so far been determined is that for the photosynthetic reaction centre [7]. Additionally, structures for bacteriorhodopsin have been determined to reasonable resolution ($\approx$3 Å) from electron microscopy [8]. These are the only two structures known to date. However, the proteins which in general are of greatest interest to the pharmaceutical and agrochemical industries are receptors. These are all membrane proteins and many of them are multimeric, such as the LGICs. The application of current molecular biology techniques has allowed many of these receptors to be cloned and the amino acid sequence to be determined (from the DNA sequence). Thus, there is a large body of sequence information available for these proteins coupled with a wealth of pharmacological and biochemical information; however, there is very little structural information and almost none at high resolution.

To attempt to provide a structural rationalization of the biochemical, pharmaceutical and biophysical information available, we have used a comparative molecular modelling approach to develop models of the nAChR agonist-binding site and the whole receptor. A fundamental aspect of this modelling has been the use of multiple amino acid sequence alignments of the LGIC subunits, of which over 50 are now known (from cloning studies) for different species and for the different agonist ligands, and which form receptors of differing ion selectivity.

Biochemical studies to localize such sites involved in agonist binding have focussed attention on two regions of the extracellular portion of the *Torpedo* nAChR. The first is the region around the paired Cys residues 192–193 of the α-subunit which are labelled by various affinity ligands, including (4-*N*-maleimide)benzyltrimethylammonium (MBTA) and bromoacetylcholine [9], after selective reduction of the disulphide bridge between them. In addition, a peptide extending from position 185 to 196 has been shown to bind α-bungarotoxin [10]. The second region is from position 125 to 147 of the α-subunit. This peptide contains a highly conserved 15-residue stretch of sequence termed the Cys-loop, so called because a disulphide bridge links Cys residues at positions 128 and 142 [9]. A synthetic peptide to this region was shown to interact with acetylcholine and α-bungarotoxin [11]. More recently, Madhok *et al.* [12] have reported that for the brain nACh receptor high-affinity binding of nicotine is specifically inhibited by antibodies raised to a peptide covering positions 3–12 of the Cys-loop of neuronal α-subunits.

The region on which we have concentrated is the Cys-loop, which is the most conserved stretch of amino acid sequence in the diverse set of sequences of LGIC subunits known to date; 4 of its 15 residue positions are invariant. In contrast, only 14 residues are invariant in the entire *N*-terminal extracellular region of LGIC subunits, which is >200 residues long. Additionally, at position 11 of the Cys-loop an invariant Asp residue occurs, which is one of only two invariant acidic residue positions present in the extracellular region of LGIC subunits and is, therefore, a good candidate for the proposed anionic site. In none of the Cys-loop sequences does an insertion or deletion of amino acid residues occur. For this reason, and because the first and last position are disulphide linked, it can be considered to be a coherent structural motif of LGIC receptors. Some earlier proposals have been made [13,14] about the structure of this region in the α-subunit of the *Torpedo* nAChR as a possible agonist-binding site, but these did not include the mutual pairing of Cys 128–142. With the establishment of the concept of a LGIC superfamily, we have been able to look beyond the nAChR and take into account the extensive biochemical, pharmacological and biophysical information for all known members of the receptor superfamily.

The Cys-Loop Model

Our modelling of the Cys-loop began with an alignment of all available Cys-loop sequences (Fig. 1*a*). From this alignment, a motif was generated which represents the conservation of amino acid residues within it (Fig. 1*b*). Using the method of Gaboriaud *et al.* [15], the α-helix and β-strand content of the loop was predicted. A marked two-residue periodicity in the average hydrophobicity defined the main-chain conformation as β-strand over positions 1–7 and over positions 10–14. A chain-reversal to allow for the formation of the disulphide bridge between the Cys residues at positions 1 and 15 is provided by a type VIa β-turn starting at position 7, and was selected by examination of similar sequence turns occurring in known protein structures [16]. This turn-type assignment is consistent with the invariance of the Pro residue at position 9 of the Cys-loop and with the occurrence of either Phe or Tyr at the preceding residue position.

The values of MacGregor *et al.* [17] obtained from an analysis of known protein structures were used to define the conformations of the side-chains. The most favoured χ_1 torsion angle, either $g+$, t or $g-$, was deduced using the alignment of Cys-loop sequences and the assigned secondary structure for each residue position. With the χ_1 torsion angles defined, the remaining side-chain torsion angles used were as given by Sutcliffe *et al.* [18] (see Table 1).

The initial model constructed was of the α_2-subunit of chick brain nAChR, the sequence of which does not include an *N*-glycosylation site at position 14, which is found in most other Cys-loop sequences. The structure was energy minimized to a maximum derivative of 0.5 kcal mol^{-1} $Å^{-1}$ (Fig. 2).

As this model is likely to be one local energy minimum structure among many, additional calculations were performed to provide further tests of the validity of this model. To determine if there were other conformations that the Cys-loop could adopt, an accessible conformation search using molecular dynamics was performed. Initial velocities of a random Maxwell–Boltzmann

(*a*)

nACh R

MUSCLE

Subunit	Species	1	2	3	4	5	6	7	8	9	10	11	12	13	14	15
Alpha	(h)	C	E	I	I	V	T	H	F	P	F	D	E	Q	N	C
	(m)	-	-	-	-	-	-	-	-	-	-	-	-	-	-	-
	(b)	-	-	-	-	-	-	-	-	-	-	-	-	-	-	-
	(c)	-	-	-	-	-	-	Y	-	-	-	-	-	-	-	-
	(x)	-	-	-	-	-	-	Y	-	-	-	-	-	-	-	-
	(t)	-	-	-	-	-	-	-	-	-	-	-	-	-	-	-
		C	E	I	I	V	T	-	F	P	F	D	E	Q	N	C
Beta	(m)	C	S	I	Q	V	T	Y	F	P	F	D	W	Q	N	C
	(b)	-	-	-	-	-	-	-	-	-	-	-	-	-	-	-
	(t)	-	T	-	K	-	M	-	-	-	-	-	-	-	-	-
		C	-	I	-	V	-	Y	F	P	F	D	W	Q	N	C
Gamma	(h)	C	S	I	S	V	T	Y	F	P	F	D	W	Q	N	C
	(m)	-	-	-	-	-	-	-	-	-	-	-	-	-	-	-
	(b)	-	P	V	-	-	-	F	-	-	-	-	-	-	-	-
	(c)	-	-	-	-	-	-	-	-	-	-	-	-	-	-	-
	(x)	-	P	V	V	-	-	-	-	-	-	-	-	-	-	-
	(t)	-	-	-	H	-	-	-	-	-	-	-	-	-	-	-
		C	-	-	-	V	T	-	F	P	F	D	W	Q	N	C
Delta	(m)	C	P	I	S	V	T	Y	F	P	F	D	W	Q	N	C
	(b)	-	-	-	-	-	-	-	-	-	-	-	-	-	-	-
	(c)	-	-	-	N	-	-	F	-	-	-	-	-	-	-	-
	(x)	-	S	-	N	-	N	-	-	-	-	-	-	-	-	-
	(t)	-	-	-	N	-	L	-	-	-	-	-	-	-	-	-
		C	-	I	-	V	-	-	F	P	F	D	W	Q	N	C
Epsilon	(b)	C	A	V	E	V	T	Y	F	P	F	D	W	Q	N	C
	(m)	-	-	-	-	-	-	-	-	-	-	-	-	-	-	-
	(r)	-	-	-	-	-	-	-	-	-	-	-	-	-	-	-
		C	A	V	E	V	T	Y	F	P	F	D	W	Q	N	C

NEURAL

Subunit	Species	1	2	3	4	5	6	7	8	9	10	11	12	13	14	15
Alpha2	(r)	C	S	I	D	V	T	F	F	P	F	D	Q	Q	N	C
	(c)	-	-	-	-	-	-	Y	-	-	-	-	-	-	-	-
		C	S	I	D	V	T	-	F	P	F	D	Q	Q	N	C
Alpha3	(r)	C	K	I	D	V	T	Y	F	P	F	D	Y	Q	N	C
	(c)	-	-	-	-	-	-	-	-	-	-	-	-	-	-	-
		C	K	I	D	V	T	Y	F	P	F	D	Y	Q	N	C
Alpha4	(r)	C	S	I	D	V	T	F	F	P	F	D	Q	Q	N	C
	(c)	-	-	-	-	-	-	-	-	-	-	-	-	-	-	-
		C	S	I	D	V	T	F	F	P	F	D	Q	Q	N	C
Beta2	(r)	C	K	I	E	V	K	H	F	P	F	D	Q	Q	N	C
		-	-	-	-	-	-	-	-	-	-	-	-	-	-	-
		C	K	I	E	V	K	H	F	P	F	D	Q	Q	N	C
Beta3	(r)	C	T	M	D	V	T	F	F	P	F	D	Q	Q	T	C
GF-nA2	(g)	C	T	M	D	V	T	F	F	P	F	D	R	Q	N	C
ARD	(d)	C	T	I	D	V	T	Y	F	P	F	D	Q	Q	T	C
ALS	(d)	C	E	I	D	V	E	Y	F	P	F	D	E	Q	T	C

Receptor	Subunit	Species	1	2	3	4	5	6	7	8	9	10	11	12	13	14	15
GABA R	Alpha1	(b)	C	P	M	H	L	E	D	F	P	M	D	A	H	A	C
	Alpha2	(b)	-	-	-	-	-	-	-	-	-	-	-	-	-	S	-
	Alpha3	(b)	-	-	-	-	-	-	-	-	-	-	-	V	-	-	-
			C	P	M	H	L	E	D	F	P	M	D	-	H	-	C
	Beta1	(r)	C	M	M	D	L	R	R	Y	P	L	D	E	Q	N	C
		(b)	-	-	-	-	-	-	-	-	-	-	-	-	-	-	-
	Beta2	(r)	-	-	-	-	-	-	-	-	-	-	-	-	-	-	-
		(b)	-	-	-	-	-	-	-	-	-	-	-	-	-	-	-
	Beta3	(r)	-	-	-	-	-	-	-	-	-	-	-	-	-	-	-
		(b)	-	-	-	-	-	-	-	-	-	-	-	-	-	-	-
			C	M	M	D	L	R	R	Y	P	L	D	E	Q	N	C
	Gamma2	(h)	C	Q	L	Q	L	H	N	F	P	M	D	E	H	S	C
Gly R	48 kD	(r)	C	P	M	D	L	K	N	F	P	M	D	V	Q	T	C

(*b*)

	1	2	3	4	5	6	7	8	9	10	11	12	13	14	15
											(−)				
Sequence Motif	C	-	h	-	h	*	-	F	P	h	D	-	Q	-	C
								Y					H		
				↑		↑		↑			↑		↑		
	+	+	+	+	t	+	+	+		+	+	t	+	+	+

Fig. 1. (*a*) *Alignment of amino acid sequences of the Cys-loop of LGIC receptors and* (*b*) *the Cys-loop sequence motif*

(*a*) The numbering of residue positions within the Cys-loop are equivalent to positions 128–142 of the α-subunit of *Torpedo* nAChR [1]. Abbreviations of species names are given as lower-case letters in parentheses: (h), human; (r) rat; (m), mouse; (b), bovine; (c), chicken; (x), *Xenopus*; (g), goldfish; (t), *Torpedo*; (d), *Drosophila.* (*b*) Invariant or strongly conserved residues are in upper-case letters. h, conserved hydrophobic; ↑, binding surface residue; (−), anionic site; *, specificity residue. The assigned χ_1 conformations (+ = *gauche*+; *t* = *trans*) used in the construction of Cys-loop models are indicated in the bottom line.

Table 1. *Initial ϕ, ψ and χ_1 angles of the Cys-loop model*

Residue	ϕ*	ψ*	χ_1†
Cys-1	−147	145	+
Ser-2	−147	145	+
Ile-3	−147	145	+
Asp-4	−147	145	+
Val-5	−147	145	*t*
Thr-6	−147	145	+
Tyr-7	−147	145	+
Phe-8	−39	132	+
Pro-9	−89	11	
Phe-10	−147	145	+
Asp-11	−147	145	+
Gln-12	−147	145	*t*
Gln-13	−147	145	+
Asn-14	−147	145	+
Cys-15	−147	145	+

* Angles are in degrees.
† + = −60, *t* = 180.

distribution for 600 K were assigned to the coordinates of the above minimized Cys-loop structure. The molecular dynamics trajectory was generated using the Verlet algorithm and a timestep of 1 fs to 5000 steps (5 ps). The analysis of the trajectory revealed no tendency for any residue to undergo major conformational change, only fluctuations around the original structure occurred. This may have been because the side-chains of the residues were limiting the conformations searched. Therefore, trajectories were generated in a similar manner for structurally modified forms of the Cys-loop (Fig. 3). Three modifications were investigated: (i) the disulphide bridge was reduced; (ii) all residues except the Cys residues substituted for Ala and (iii) the disulphide bridge was reduced when all other residues were Ala. In the first two cases (i and ii), there was little change in the main-chain conformation. Only when the disulphide bridge was reduced with the all Ala Cys-loop (iii) did major conformational changes occur to the Cys-loop structure. This analysis indicated that the residue side-chains as well as the disulphide bridge of the Cys-loop act to constrain its main-chain flexibility. In addition, a high degree

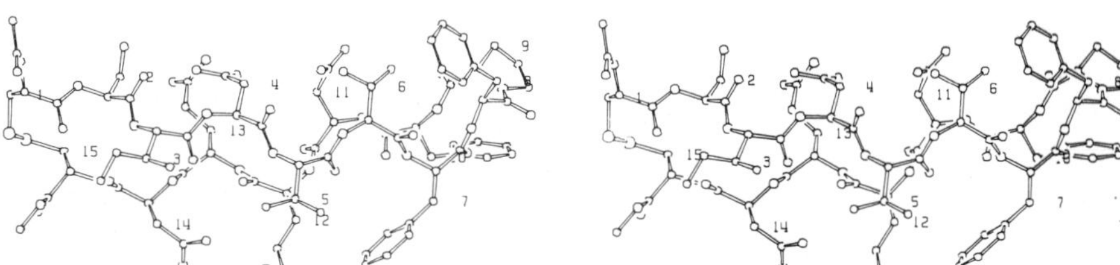

Fig. 2. *Stereo view of energy minimized model of the Cys-loop structure of the α_2-subunit of chick brain nAChR*

Numbering refers to the position of residues within the Cys-loop. Hydrogen atoms are not shown.

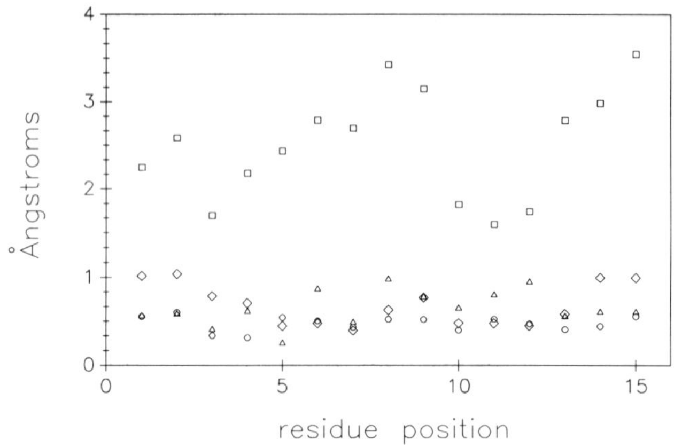

Fig. 3. *Standard deviation of C-α atom coordinates from the average coordinate set in the dynamic trajectories of the Cys-loop of the* α_2*-subunit of chick brain nAChR and modified forms of it*

○, The Cys-loop; △, Ala-substituted Cys-loop; ◇, the Cys-loop with the disulphide bridge reduced; □, Ala-substituted Cys-loop with the disulphide bridge reduced.

of rigidity for the Cys-loop is suggested by the absence of Gly residues in any of the known Cys-loop sequences and the common occurrence at positions 3 and 5 of β-branched residues which enables the side-chains to interlock.

The next phase of the modelling was to incorporate information from the Cys-loop sequences of the other members of the LGIC superfamily. These models were constructed by mutating the residues of the initial model to each of the other Cys-loop sequences in turn, followed by energy minimization of each Cys-loop. None of these minimized structures had any major deviations from the β-hairpin with a type VIa turn motif.

The comparison of the different Cys-loop models revealed a marked conservation in the amino acid groups surrounding the invariant Asp residue at position 11, on the hydrophilic face. An invariant Pro at position 9 appears to have a structural role in maintaining the stereochemistry of the type VIa turn. A Phe residue at position 8 occurs in all LGIC sequences, except the β-subunit of the $GABA_A$ receptor, in which case a Tyr residue is present (see Fig. 1*a* and discussion below). Only Glu or His occur at position 13, indicating a requirement for a hydrogen-bond donor at this position. As discussed below, the residue at position 6 is proposed as conferring selectivity in the recognition of different LGIC receptor agonists. Although the residue at this position does vary between members of the superfamily, it is highly conserved for a given LGIC subunit type (see Fig. 1*a*).

On the hydrophobic face, the residues at positions 3, 5, and 10 are invariably hydrophobic and a patch is formed that could contribute to the inner-core of the folded protein. An Asn residue often occurs at position 14 of this same face as part of an *N*-glycosylation consensus sequence. Experimental evidence indicates that in the *Torpedo* nAChR the site is glycosylated [19,20]. It is, therefore, likely that this site is at the protein surface.

Although a number of Cys-loops are found in the known protein structures for residue separations of 14 and 16 residues, very few 15 residue loops occur

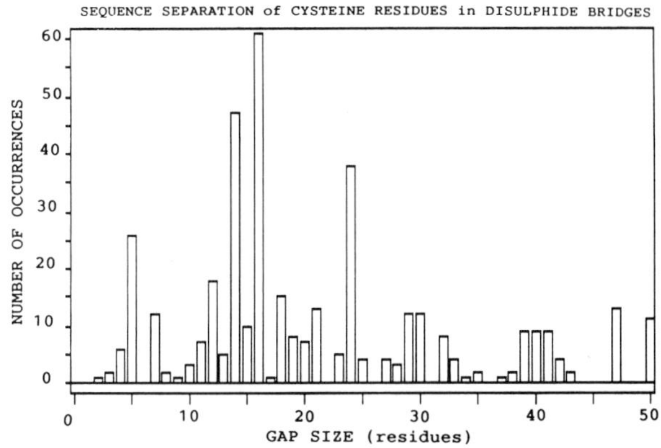

Fig. 4. *Histogram of disulphide-bridged Cys loops that occur in the Brookhaven database by number of residues in the loop, produced by a BIPED [49] query*

The number of residues in the loop includes the Cys residues.

(Fig. 4). This is likely to be due to the necessity to distort the β-sheet structure to form the disulphide bridge, the hydrogen-bonded residue pairs in standard β-hairpin structures being separated by an even number of residues. This requirement for a distortion (ϕ −80, ψ 100 in the chick brain α_2-subunit model) may select for Pro at position 2 of the Cys-loop, which is found in 10 of the 46 Cys-loop sequences in Fig. 1(*a*). Interestingly, in the Cys-loop models with Ser and Thr at this position (15 of the 46) the same main-chain distortion was stabilized by the formation of a side-chain to main-chain hydrogen bond.

Modelling the Ligand–Receptor Interactions

To incorporate the three-dimensional information inherent in the structure of agonists and antagonists known to interact with the various LGIC agonist-binding sites, we have modelled the binding-site complex of a number of these agonists. This 'docking' synthesized the sequence information on the groups surrounding the invariant Asp (which could not have been assessed from inspection of the linear sequence alone), together with the common structural and electronic features of the agonists. The structural and conformational requirements for agonist binding to the receptors for ACh [21], GABA [22] and Gly [23], based on Quantitative Structure–Activity Relations (QSAR) studies of each of them separately, have been extensively covered in the literature. Using the idea of a proto-binding site which evolved for differential specificity in the different LGIC members, we have suggested a unified model of binding for all LGIC receptor agonists.

This is based on the following model of the general pharmacophore for all LGIC ligands. It consists of a positively charged centre (termed the 'positive pole'), which is essential, and a π-electron system containing an sp^2 hybridized electronegative centre that produces a local dipole in the π-electron system. The distance between the nitrogen atom of the positive pole and the electronegative atom is 4.5–5.5 Å for ACh and GABA ligands, whereas for Gly the distance is around 3.5 Å (Fig. 5).

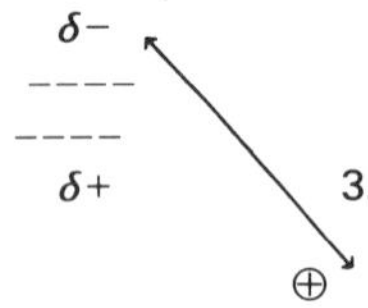

Fig. 5. *Pharmacophore model for LGIC agonists, showing the structural relations between the 'positive pole' and the electronegative sp^2 hybridized system*

Fig. 6 shows one example of our modelling of the binding-site complexes for nicotine docking to the α_2-subunit of the chick brain nAChR. Initially, (−)nicotine in its pharmacophore conformation was docked on to the Cys-loop model of the α_2-subunit of chick brain nACh receptor graphically, using the binding model described above to guide the positioning of the ligand. This docking model was then energy minimized to a maximum derivative of 0.5 kcal mol^{-1} $Å^{-1}$. The two most important interactions that were considered in this binding model, and which have been proposed as common recognition features of all LGIC agonists by their receptors, are as follows: first, there is an ion-pair interaction between the protonated tertiary amine (the positive pole) of (−)nicotine and the invariant Asp-11; secondly, there is the interaction of the protons around the ε-carbon and the conserved aromatic, Phe-8 in this case, with the nitrogen atom region of the pyridyl ring of nicotine, this heterocylic ring being the sp^2 hybridized electronegative centre for nicotine. The interaction is one in which the plane of the aromatic ring is orthogonal to that of the agonist π-electron system (such types of interaction have been documented in the literature [24–28]). In addition, the conserved local dipole of the agonist π-electron system is favourably oriented in the electrostatic field of the invariant Asp group.

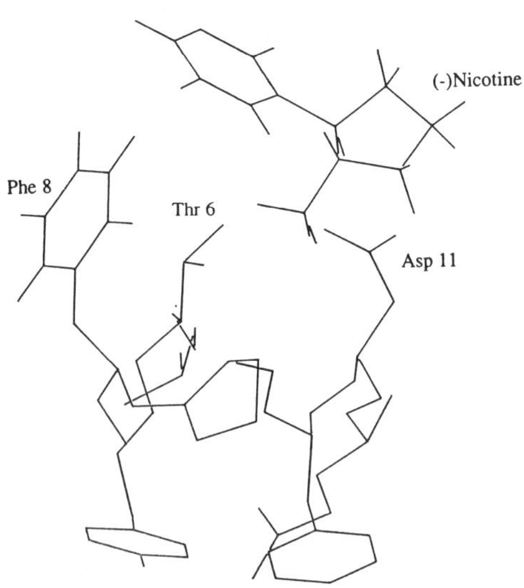

Fig. 6. *Energy minimized model of (−)nicotine docked onto the Cys-loop of the α_2-subunit of chick brain nAChR*

Table 2. *Proposed Cys-loop residues affected by chemical modification and observed effects on ligand binding*

Chemical reagent	Cys-loop residues modified	LGIC subunit	Experimental effect
2,3-Butadione [44], Phenylglyoxal [44]	Arg-6	GABA β	GABA binding reduced
p-Diazobenzenesulphonic acid [45], Tetranitromethane [45], *N*-acetylimidazole [45]	Tyr-8	GABA β	GABA binding reduced
Diethylpyrocarbonate [45, 46]	His-6 His-13	GABA γ^2	Benzodiazepine binding reduced
	His-4 His-13	GABA α	GABA binding unchanged
Fluorescein isothiocyanate [47, 48]	Lys-6	Gly 48 kDa	Gly binding reduced

We have proposed that the residue at position 6 of the Cys-loop is a key determinant of the selective recognition of GABA, Gly and ACh by their receptors. This residue is changed in the different receptor families and changes in subunits of the same receptor that interact with different ligands. In the β-subunit of the $GABA_A$ receptor, it is an Arg residue, it is a Lys residue in the 48 kDa subunit of the Gly receptor and a Thr residue in the α-subunit of the nAChR. The different size, charge and hydrogen-bonding capacity of these residues can be associated with a specific interaction with each of the different ligands that bind to that receptor, thus providing a method of discriminating between the various ligands.

Experimental Support for the Docking Model

The major strength of the modelling is that, first, a single type of structure accommodates all of the sequence variations of the Cys-loop in the >50 known polypeptides of the LGIC superfamily. Secondly, specific chemical modifications of the $GABA_A$ receptor and Gly receptor (Table 2) can be accounted for by the residues that are spatial neighbours of the invariant Asp group on the hydrophilic face of the Cys-loop, and in particular the residue at position 6 of the Cys-loop.

Extended Binding-Site Model

The Cys-loop structure is clearly a model for only a part of the complete ligand-binding site for the LGIC receptors, as it does not provide for a binding-site pocket, but a binding-site surface. The complete ligand-binding site will require additional parts of the *N*-terminal sequence to surround the binding-site face of the Cys-loop to provide some form of binding-site pocket, but leaving the binding-site face of the Cys-loop exposed to solvent. An immediate candidate for these additional parts is the region of the sequence around Cys 192–193.

From studies using reagents specific for the binding site that alkylate Cys 192–193 of the *Torpedo* nAChR, it is often suggested that this region plays a dominant role in agonist binding. However, for such reagents the active part of the molecule is in a position equivalent to the bromomethyl group of bromoacetylcholine and is, therefore, probing for residues in a limited region of space around the binding site. Furthermore, Cys 192–193 are present only in the α-subunits of the nAChR and do not occur in any other LGIC subunit known to date. Also, the surrounding region even between α-subunits of the nAChR of the muscle/electric organ type [29] and the neuronal subtypes of nACh receptor [30] can be seen to have accepted residue substitutions and insertions/deletions. Moreover, Cys 192–193 have been experimentally mutated to Ser without complete loss of agonist binding [31]. This noted lack of structural conservation is in our view inconsistent with the overlap in nicotinoid pharmacology seen for members of the nicotinic family of receptors [32]. Therefore, we believe this region is close to the agonist-binding site as part of a binding cleft and that it may contain important determinants of toxin binding, but it may not necessarily be the primary determinant responsible for specifying agonist binding.

Additional information on the binding-site cleft can be provided by pharmacophore mapping using the extensive range of agonists and antagonists known for the LGIC family. Using the pharmacophore model, the following agonists and antagonists were superimposed using the three points, the 'positive pole' atom, the electronegative and electropositive atom centres of the dipole of the π-electron system (Table 3). The superimposition was viewed graphically and the spatial requirements for the binding-site pocket determined (Fig. 7). Interestingly, the agonists can be seen to be small molecules compared with the antagonists, which occupy a much larger volume of space.

Combining these two sources of information, active-site labelling studies and pharmacophore mapping, we have extended the original Cys-loop binding site model to produce a model of the binding-site pocket (Fig. 8).

Table 3. *Atoms used in superimposition of agonists and antagonists*

THIP, 4,5,6,7-tetrahydroisoxazolo[5,4-*c*]pyridin-3-ol.

Molecule	'Positive pole' atom	Electronegative atom	Electropositive atom
Agonists:			
Cytisine	*N* of secondary amine	*O* of carbonyl of 2-pyridone	*C* of carbonyl of 2-pyridone
THIP	*N* of secondary amine	*O* of carbonyl of isoxazolin-3-one	*C* of carbonyl of isoxazolin-3-one
Antagonists:			
Methyllycaconitine	*N* of tertiary amine	*O* of carbonyl of ester side-chain	*C* of carbonyl of ester side-chain
Bicuculline	*N* of tertiary amine	*O* of carbonyl of γ-lactone	*C* of carbonyl of γ-lactone

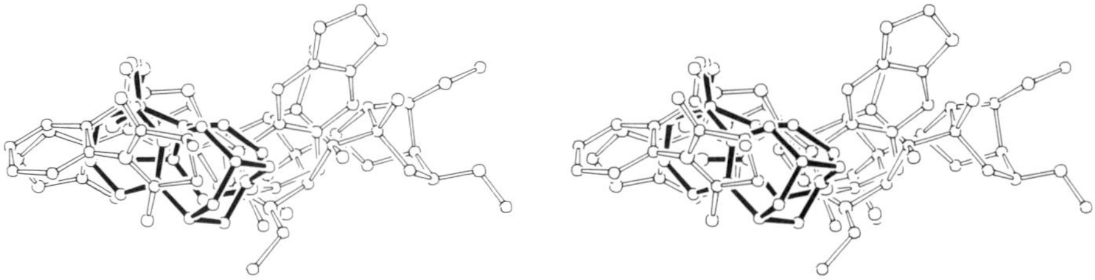

Fig. 7. *Stereo view of the superimposition of agonists, cytisine and THIP, and antagonists, methyllycaconitine and bicuculline*

The agonists have solid bonds. Hydrogen atoms are not shown.

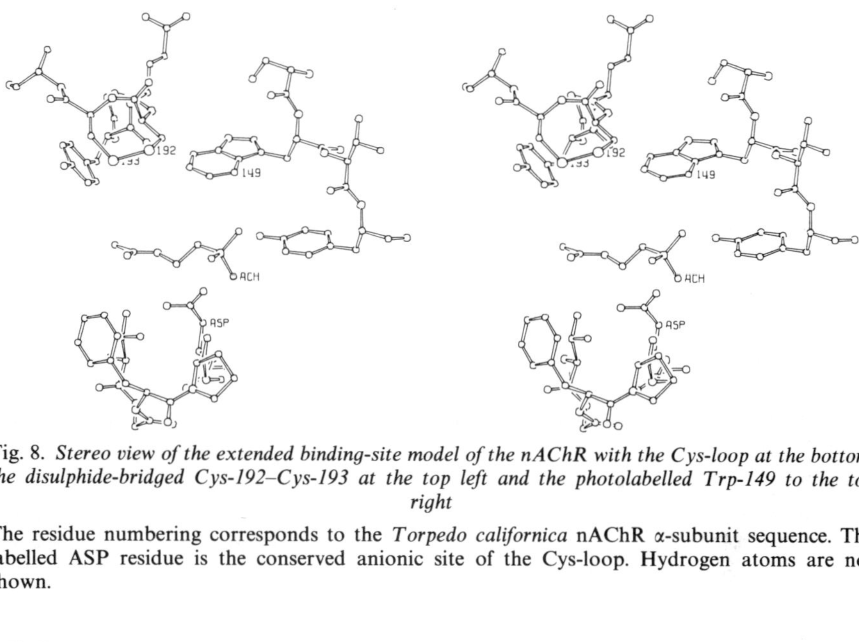

Fig. 8. *Stereo view of the extended binding-site model of the nAChR with the Cys-loop at the bottom, the disulphide-bridged Cys-192–Cys-193 at the top left and the photolabelled Trp-149 to the top right*

The residue numbering corresponds to the *Torpedo californica* nAChR α-subunit sequence. The labelled ASP residue is the conserved anionic site of the Cys-loop. Hydrogen atoms are not shown.

Whole Receptor Modelling

The higher level phase of the modelling was to model the rest of the receptor, including the transmembrane helices and the complete extracellular domain.

Transmembrane Helices Modelling

There are four regions (M1–M4) that have been identified as putative transmembrane helices based on hydrophobicity profiles [33], and one region (MA) that has been identified as a highly charged amphiphilic helix [24]. We generated helices for these five regions using a standard helix geometry of -65, -40 and analysed the side-chain patterns visually. The amphiphilic helix has all the charged residue side-chains aligned down one side of the helix surface and because of this it has been suggested that it is the helix which lines the ion-channel [34,35]. Recent work by Numa and co-workers [31,36] demonstrated that partial ion-channel activity still existed even after the deletion (by molecular biological techniques) of the MA helix region. Attention now has focused on the M2 helix as the ion-channel lining helix. This contains a number of Ser and Thr residues which can be lined up along one side of the helix surface. Also, novel peptides have been synthesized based on the M2 sequence which have been shown to act as ion-channels [37]. Using the known structure of myohemerythrin, a four-helix anti-parallel bundle protein with helices of a length that could span the membrane, we have generated a model for the transmembrane helices by superimposing the four helices M1–M4 on to the known helices. The approach used at the superimposition step was to inspect the sequence patterns in an alignment of myohemerythrin sequences and to fit the pattern seen in the LGIC sequences on to this. Considerations included conservation, variability, hydrophobicity and size, as well as preference of amino acids to occur at the helix termini [38]. In addition,

Fig. 9. *View looking from the cytoplasmic side of the receptor through the ion-channel of the model of the transmembrane region*

The helices M1–M4 of one subunit are labelled, the helices in the other subunits are in the same relative positions, M2 always being the central helix. Only Cα atoms are shown.

biochemical data assigning residue positions of M2 as sites lining the ion-channel were used [39]. Five copies of this model were docked together with, in each case, M2 forming the central ion-channel lining helix to model the complete receptor ion channel (Fig. 9). M1 is the most tightly packed helix, making both intra-subunit contacts (M2 and M4), and inter-subunit contacts (M2 and M3 of next subunit). This leads to the M4 helix lying on the outside of the helix bundles and making no contacts with the helices of adjacent sub-units. Thus, from the model there seem to be few structural constraints on this helix and the sequence alignments show that this is the least conserved helix. Further, Numa and co-workers have replaced this helix with a hydrophobic helix from insulin and retained activity [36], which again indicates there are few structural constraints on this helix.

Extracellular Domain Modelling

The first assumption used in modelling the extracellular domain is that as it appears to be entirely external to the membrane, it is highly likely that its structure is determined by the same rules as for normal globular proteins, for which we have a large body of structural data. However, using database-searching techniques based on sequence information alone, little homology was found to any known structure (V. B. Cockcroft & D. J. Osguthorpe, unpublished work).

Recently, we have developed a novel method for scanning a database of known structures using a probe sequence based on aligned sequences of unknown structure (V. B. Cockcroft & D. J. Osguthorpe, unpublished work). This method requires having available a set of sequences for the unknown structure which are around 30–50% homologous, so that a multiple sequence alignment algorithm can produce a good alignment yet not so similar that too few amino acids are substituted. This alignment of sequences is used to produce a predicted surface accessibility at each amino acid position in the sequence (based on a scale of accessibilities defined for each amino acid). These accessibilities are then used to scan the computed accessibilities of known structures in the database to find segments of known structures which correspond in accessibility to the unknown structure.

A subset of the LGIC sequences fulfilling the above requirements was selected and the method has been applied to them. This resulted in the protein PPase being identified as having a high degree of similarity to the extracellular domain of the nAChR. A pair-wise alignment of these two sequences gave an 18% identity conservation between them (Fig. 10). This low level of sequence homology has in other modelling work [42, 43] been used to generate structures by homology modelling which have been shown to be correct overall by subsequent X-ray studies.

Thus, we have used this alignment of the nAChR and the known PPase structure to generate a framework model for the extracellular domain of the receptor by amino acid mutation of the PPase structure. Analysis of the model so far indicates agreement with the biochemical data. The MIR occurs on the surface of this protein as does the Cys-loop.

```
   >>>>>>> prepeptide >>>>>>>>
1. MEPRP-------LLLLLGLCSAGLVLGSEHETRLVA-KLFEDYNSVVRPVEDHRQAVEVTVG
2. TYTTRQIGAKNTLEYKVYIEKDGKPVSAFHDIPLYADKEDNIFNMVVEIPRWTNAKLEITKE
              #   ~ ~   #  ~~~ #~  # # #  ~ ~# ##        ~#~#

1. LQLIQLINVDEVNQIVTTNVRLKQQWVD-YNLKW----NPDDYGGVKKIHIPSEKIWRPDLV
2. ETLNPIIQNTKGKLRFVRNCFPHHGYIHNYGAFPQTWEDPNVSHPETKAVGDNNPIDVLQIG
     #  ~#~    ~    #    ~ ~~~ #  ~     ~#~     #    ~~ #   ~~

                                      <   CYS-LOOP  >
1. LYNNADGDFAIVKFTKVL-LDYTGHITWTPPAIFKSYCEIIVTHFPFDEQNCSMKLGTWTYD
2. ETIAYTGQVKEVKALGIMALLDEGETDWKVIAIDINDPLAPKLNDIEDVEKYFPGLLRATDE
         #~  ##   ~~ #   #~  #   ##  ~      ~   #  ~~    #   #  ~

                                            <192-193>
1. GSVVVINPESDQPDLSNFMESGEW------VIKESR-GWKHWVFYACC------------PS
2. WFRIYKIPDGKPENQFAFSGEAKNKKYALDIIKETHNSWKQLIAGKSSDSKGIDLTNVTLPD
      ~   #~~  ~   #   ~        ~###~~ ~##~ ~                #

1. TP-YLDITYHFVMQRLP
2. TPTYSKAASDAIPPASP
   ## #   ~ ~ ~    #

# = identity
~ = conserved (PTSAG,MLIV,QEND,RKH,WYF,C)
```

Fig. 10. *Pairwise alignment of yeast PPase with the N-terminal sequence of the bovine nAChR* The sequence labelled 1 is the nAChR and sequence 2 is PPase

Conclusions

The main benefit of the construction of explicit atomic models, besides giving insights into receptor structure and function, is that it allows specific experiments to be designed and prioritized to test the model, and to comprehend and verify current experimental data. Another potential use of such models is to design and test ligands of pharmacological and therapeutic importance. We believe the modelling in this particular area will prove itself to be of great value in producing safer, more effective drugs and in inventing completely novel drugs, in conjunction with the more classical approaches of pharmacophore modelling.

The molecular modelling is presented as a first step in the understanding of the structure–activity relationships of receptor ligands, while taking into consideration the receptor sites themselves. Finally, the comparative approach to receptor structural modelling we have adopted has much potential value in the rational design of novel ligands with subtype and species specificity.

We thank the S.E.R.C. and Shell Research (U.K.) Ltd for financial support. D.J.O. would also like to acknowledge the Nuffield Foundation for providing the funds for the plotter used to produce the Figures.

References

1. Noda, M., Takahashi, H., Tanabe, T., Toyosato, M., Furutani, Y., Hirose, T., Asai, M., Inayama, S., Miyata, T. & Numa, S. (1982) *Nature* (*London*) **299**, 793–797
2. Schofield, P. R., Darlison, M. G., Fujita, N., Burt, D., Stephenson, F. A., Rodriguez, H., Rhee, L. M., Ramachandran, J., Reale, V., Glencorse, T. A., Seeburg, P. H. & Barnard, E. A. (1987) *Nature* (*London*) **328**, 221–227
3. Grenningloh, G., Rienitz, A., Schmitt, B., Methfessel, C., Beyrether, K., Gundelfinger, E. D. & Betz, H. (1987) *Nature* (*London*) **328**, 215–220
4. Toyoshima, C. & Unwin, N. (1988) *Nature* (*London*) **336**, 247–250
5. Unwin, N. (1989) *Neuron* **3**, 655–676
6. Mitra, M., McCarthy, M. P. & Stroud, R. M. (1989) *J. Cell Biol.* **109**, 755–774
7. Deisenhofer, J., Epp, O., Mikki, K., Huber, R. & Michel, H. (1985) *Nature* (*London*) **318**, 618–624

8. Henderson, R., Baldwin, J. M., Ceska, T. A., Zemlin, F., Beckmann, E. & Downing, K. H. (1990) *J. Mol. Biol.* **213**, 899–929
9. Kao, P. N. & Karlin, A. (1986) *J. Biol. Chem.* **261**, 8085–8088
10. Neumann, D., Barchan, D., Fridkin, M. & Fuchs, S. (1986) *Proc. Natl. Acad. Sci. U.S.A.* **83**, 9250–9253
11. McCormick, D. J. & Atassi, M. Z. (1984) *Biochem. J.* **224**, 995–1000
12. Madhok, T. C., Chao, C. C., Matta, S., Hong, A. & Sharp, B. M. (1989) *Biochem. Biophys. Res. Commun.* **165**, 151–157
13. Smart, L., Meyers, H., Hilgenfeld, R., Saenger, W. & Maelicke, A. (1984) *FEBS Lett.* **178**, 64–68
14. Luyten, W. H. M. L. (1986) *J. Neurosci. Res.* **16**, 51–73
15. Gaboriaud, C., Bissery, V., Benchetrit, T. & Mornon, J. P. (1987) *FEBS Lett.* **224**, 149–155
16. Chou, P. Y. & Fasman, G. D. (1977) *J. Mol. Biol.* **115**, 135–175
17. MacGregor, M. J., Islam, S. A. & Sternberg, M. J. E. (1987) *J. Mol. Biol.* **198**, 295–310
18. Sutcliffe, M. J., Hayes, F. R. F. & Blundell, T. L. (1987) *Protein Eng.* **1**, 385–392
19. Conti-Tronconi, B. M., Hunkapiller, M. W. & Raftery, M. A. (1984) *Proc. Natl. Acad. Sci. U.S.A.* **81**, 2631–2634
20. Nomoto, H., Takahashi, N., Nagaki, Y., Endo, S., Arata, Y. & Hayashi, K. (1986) *Eur. J. Biochem.* **57**, 233–242
21. Sheridan, R. P., Nilakantan, R., Dixon, J. S. & Venkataraghavan, R. (1986) *J. Med. Chem.* **29**, 899–906
22. Krogsgaard-Larsen, P., Nielsen, L., Falch, E. & Curtis, D. R. (1985) *J. Med. Chem.* **28**, 1612–1617
23. Tokutomi, N., Kaneda, M. & Akaike, N. (1989) *Br. J. Pharmacol.* **97**, 353–360
24. Thomas, K. A., Smith, G. M., Thomas, T. B. & Feldmann, R. J. (1982) *Proc. Natl. Acad. Sci. U.S.A.* **79**, 4843–4847
25. Czupinski, O., Malarski, Z. & Sobczyk, L. (1980) *Bull. Acad. Pol. Sci. Sev. Sci. Chim.* **28**, 661–667
26. Burnley, S. K. & Petsko, G. A. (1988) *Adv. Protein Chem.* **39**, 125–189
27. Pettersen, I. & Liljefors, T. (1987) *J. Comp. Chem.* **8**, 1139–1145
28. Matthews, D. A., Alden, R., Bolin, J. T., Freer, S. T., Hamlin, R., Xuong, N., Kraut, J., Poe, M., Williams, M. & Hoogstein, K. (1977) *Science* **197**, 452–455
29. Kubo, T., Noda, M., Takai, T., Tanabe, T., Kayano, T., Shimizu, S., Tanaka, K., Takahashi, H., Hirose, T., Inayama, S., Kikuno, R., Miyata, T. & Numa, S. (1985) *Eur. J. Biochem.* **149**, 5–13
30. Wada, K., Ballivet, M., Boulter, J., Connolly, J., Wada, E., Deneris, E. S., Swanson, L. W. & Heinemann, S. (1988) *Science* **240**, 330–334
31. Mishina, M., Tobimatsu, T., Imoto, K., Tanaka, K., Fulita, Y., Fukuda, Y., Hirose, T., Inayama, S., Takahashi, T., Kuno, M. & Numa, S. (1985) *Nature (London)* **313**, 364–369
32. Wonnacott, S. (1986) *J. Neurochem.* **47**, 1706–1712
33. Claudio, T., Ballivet, M., Patrick, J. & Heinemann, S. (1983) *Proc. Natl. Acad. Sci. U.S.A.* **80**, 1111–1115
34. Finer-Moore, J. & Stroud, R. M. (1984) *Proc. Natl. Acad. Sci. U.S.A.* **81**, 155–159
35. Stroud, R. M. & Finer-Moore, J. (1985) *Annu. Rev. Cell Biol.* **1**, 317–351
36. Tobimatsu, T., Fujia, Y., Fukuda, K., Tanaka, K., Mori, Y., Konno, T., Mishina, M. & Numa, S. (1987) *FEBS. Lett.* **222**, 56–62
37. Lear, J. D., Wasserman, Z. R. & DeGrado, W. F. (1988) *Science* **240**, 1177–1181
38. Richardson, J. S. & Richardson, D. C. (1988) *Science* **240**, 1648–1652
39. Guy, H. R. & Hucho, F. (1987) *Trends Neurosci.* **10**, 318–321
40. Reference deleted
41. Reference deleted
42. Pearl, L. H. & Taylor, W. R. (1987) *Nature (London)* **329**, 351–354
43. Blundell, T. & Pearl, L. (1989) *Nature (London)* **337**, 596–597
44. Widdows, K. B., Kirkness, E. F. & Turner, A. J. (1987) *FEBS Lett.* **222**, 125–128
45. Maksay, G. & Ticku, M. K. (1984) *J. Neurochem.* **42**, 1715–1727
46. Lambolez, B. & Rossier, J. (1987) *FEBS Lett.* **219**, 301–305
47. Gomez, A. R., Fernandez-Shaw, C., Valdivieso, F. & Mayor, Jr., F. (1989) *Biochem. Biophys. Res. Commun.* **160**, 374–381
48. Pritchett, D. B., Sontheimer, H., Shivers, B. D., Ymer, S., Kettenmann, H., Schofield, P. R. & Seeburg, P. H. (1989) *Nature (London)* **338**, 582–585
49. Islam, S. A. & Sternberg, M. J. E. (1989) *Protein Eng.* **2**, 431–442

Biochem. Soc. Symp. **57**, 81–90
Printed in Great Britain

Monoamine Neurotransmitter Receptors: Ligand–Receptor Models

JOHN SAUNDERS*‡ and JOHN B. C. FINDLAY†

**Merck, Sharpe & Dohme Research Laboratories, Neuroscience Research Centre, Terlings Park, Eastwick Road, Harlow, Essex CM20 2QR, U.K. and †Department of Biochemistry, University of Leeds, Leeds LS2 9JT, U.K.*

Synopsis

Recent medicinal chemistry studies have led to the idea that muscarinic ligands of differing efficacy employ distinctly different pharmacophores when interacting with their specific receptor. Although the primary recognition element, the cationic head group, remains constant for all ligands, two H-bonding interactions are used selectively to stabilize agonist binding. In contrast, antagonists rely predominantly upon hydrophobic binding in the vicinity of the acetylmethyl group present in the endogenous transmitter. These ideas have been evaluated in the context of a three-dimensional model of the muscarinic receptor derived originally from bacteriorhodopsin. Together these studies have allowed speculations to be made regarding the sequence of events which lead to coupling to G-protein. Although an Asp residue on helix III of the seven-helical model is known to be involved as the primary recognition site for all ligands, a second, more deeply buried Asp on helix II is suggested to represent the ultimate agonist-binding site from which receptor activation is triggered. Similar arguments are also applied to homologous receptor systems, in particular the *β*-adrenergic receptor.

Introduction

Work over the last 4 years was aimed at designing full agonists for the cortical M_1 muscarinic receptor (MR) which may be useful in the understanding of cholinergic mechanisms underlying Alzheimer's disease. During the course of this work, an interest was developed in understanding the structure of the target protein itself to rationalize the structure–activity relationships evolving from the chemistry. This paper will review data already published in this area [1] and also expand the coverage to include information emerging from similar studies on the *β*-adrenergic receptor [2]. The question that will be addressed is: can the current three-dimensional models for monoamine receptors linked to G-proteins assist in both our understanding of ligand-induced receptor activation and the design of new agonists for such receptors? The message that will be conveyed is that the models are far from good

‡ Present address: Medicinal Chemistry Department II, Glaxo Group Research Ltd, Greenford Road, Greenford, Middlesex UB6 0HE, U.K.

enough to attempt to explain such things as the subtleties which distinguish the various receptor subtypes, but have advanced to the level of allowing reasonable speculations to be made regarding ligand binding and receptor activation. Just one part of the signal transduction cascade will be discussed: the formation of the activated ligand–receptor complex from ground-state receptor and free agonist.

Derivation of the Receptor Model

Over the past 3 or 4 years there has been a rapid advance in our understanding of G-protein-linked receptors, primarily because of information derived from molecular biological studies [3]. Sequence and structural information first became available for rhodopsin and bacteriorhodopsin. The current model for the MR has evolved from the very limited structural data available for bacteriorhodopsin [4] in a low-resolution (6 Å) electron diffraction study from which the protein has been assumed to consist of seven helical arrays which successively traverse the lipid membrane of the cell wall. Chemical and proteolytic studies, primarily on rhodopsin, have helped pinpoint the buried regions and have shown the *N*-terminal sequence to be extracellular. Findlay's group at Leeds were able to map the α-carbon chain of rhodopsin on to the electron density map of bacteriorhodopsin [5]. When the amino acid sequence for the M_1 receptor became known an homology model was constructed and it is this model which will be described here.

Receptor Structure and Putative Functional Residues

A complete description of the M_1 receptor model will be given elsewhere; an outline of the backbone atoms are shown (Fig. 1) with the extracellular surface, and hence the *N*-terminus, towards the top. The 150–250 residue loop present in the various subtypes of MR and which connects the fifth and sixth helices has been deleted simply because we do not know how to model this region, it being less conspicuous (16 residues) in bacteriorhodopsin. Since it is now well established that the ligand-binding site is within the membrane-spanning regions (see below), this deletion is of no consequence to the present argument, although the 5–6 loop is critically implicated in the formation of a functional complex with G-protein [6]. As can be clearly observed from the original model of bacteriorhodopsin, each of the helical domains is far from perfect with all but the fourth having at least one major disruption to helical structure. In the MR, these positions of deviation are often associated with the presence of Pro residues and lead to the speculation that they may represent sites of conformational lability which is assumed to be necessary for receptor activation. Within the putative helical domains of the model, there are four acidic residues conserved throughout the family of monoamine neutrotransmitter receptors and these are highlighted with a dot surface.

In the *β*-receptor, each of the four conserved acidic residues within the helical domains has been mutated separately to identify the site or sites involved in the primary Coulombic interaction with ligand [2]. Glu-107,

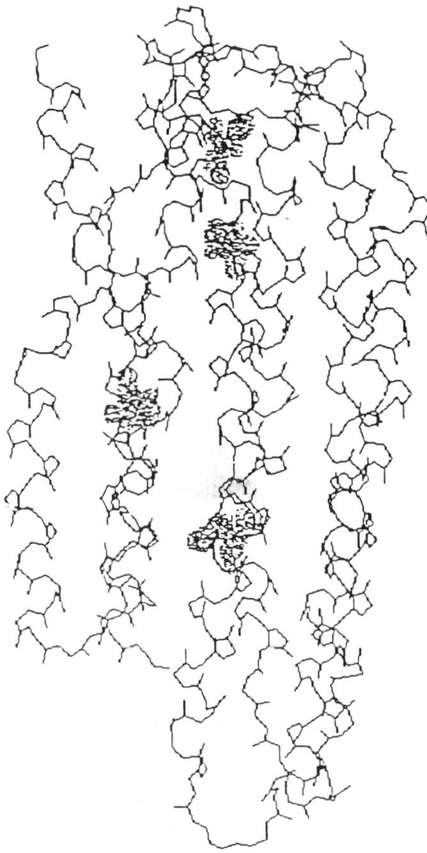

Fig. 1. *Model of the muscarinic receptor*
Except for the functional residues (highlighted), only the backbone atoms are shown for clarity.

located near to the surface of the receptor on helix III, does not seem to be involved in ligand binding, since agonists and antagonists show similar affinity to wild-type. However, a second acid residue, Asp-113 also on helix III, but buried by about 10 Å, is vital for both agonist and antagonist binding as witnessed by the 10000-fold reduction in affinity for the Asn-113 mutant. This residue is seen as the primary recognition site for all ligands. However, by using very high concentrations of agonist, it was possible to show that the Asn-113 mutant is still fully coupled to cyclase suggesting the possibility that a second, more deeply buried Asp residue may function to bind agonist ligands. This is indeed the case and the involvement of Asp-79 on helix II some 15 Å from the receptor surface leads to the hypothesis that agonists but not antagonists are able to equilibrate between Asp-113 and Asp-79 and, furthermore, receptor activation demands agonist occupancy at this latter site.

Where comparable data are available, similar observations have been recorded for the M_1 receptor [7]. In the MR, the uppermost acidic residue is Asp-99 (equivalent to Glu-107 in the β-receptor) and, as with the β-receptor, this site does not seem to be involved, there being only a small decrease in ligand affinity for the Asn-99 protein. Regrettably, agonist-binding data for the Asn-105 mutant were not reported and its coupling status to phosphatidylinositol (PtdIns) remains undetermined; one would predict that the receptor will still be fully coupled, albeit at high agonist concentrations. Finally, the Asp located on helix II, Asp-71 (Asp-79 in the β-receptor), also influences coupling, since the PI response to carbachol was almost completely abolished over the normal concentration range in the Asn-71 mutant.

Properties Influencing Ligand Efficacy at the MR

In cerebral cortex, those muscarinic receptors which are coupled positively to the turnover of PtdIns hydrolysis appear to have only very limited receptor reserve and, therefore, this receptor system can be used to measure efficacy of various muscarinic agonists critically. It is now known that there are at least three receptor subtypes: M_1, M_3 and M_5, which are capable of positively stimulating PtdIns hydrolysis [8] and binding data in cortical slices must be interpreted with this in mind. Is there a relatively simple method of determining efficacy at the cortical site? It turns out that a simple two-step radioligand-binding assay developed by Freedman at Merck [9] provides the answer, and simultaneously gives a measure of the affinity of a given ligand for both the low- and high-affinity states of the receptor. Radiolabelled *N*-methylscopolamine (NMS) is used to label selectively the low-affinity state of the receptor, while radiolabelled oxotremorine-M (OXO-M) labels the high-affinity state. Only when these types of data are available can one start speculating on which pharmacophoric elements distinguish agonists from antagonists. Later, a rationalization will be presented for exactly how this assay does indeed give a measure of efficacy, but for now it can be assumed that the binding ratio, the NMS/OXO-M ratio, is predictive of efficacy. The logarithm of this ratio has been shown to correlate with the ability of the ligand to stimulate the hydrolysis of cortical PtdIns [9]. Four broad categories of ligand can be defined according to their efficacy as estimated from this ratio: antagonists show equal affinity in both binding assays and thus have ratios close to unity; weak partial agonists have low ratios between 10 and 200; partial agonists display intermediate ratios in the range 200–800, whereas, at the other end of the continuum, full agonists display ratios in excess of 800. In the course of studies aimed at achieving high-efficacy MR agonists devoid of a quaternary ammonium entity, it has been possible to design antagonists, partial agonists and full agonists with affinity, and, in the case of the agonists, efficacy comparable to or greater than that of the classical quaternary ammonium ligand [10]. Thus, from a single series of oxadiazole-based ligands, three representative examples have been selected which have been used in the ligand-docking experiments (Fig. 2).

The first oxadiazole has a diphenylhydroxy substituent reminiscent of that in the classical muscarinic antagonist, quinuclidinyl benzilate, it has subnanomolar affinity and, with a ratio close to 1, zero efficacy. It is therefore a potent antagonist at the MR. As we move to a methyl substituent, affinity for both affinity states of the receptor is diminished through loss of lipophilic

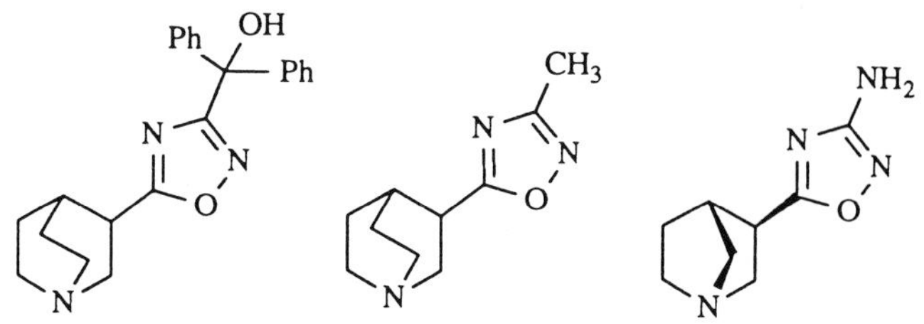

NMS[a]	0.00012	0.44	0.031
OXO-M[b]	0.000067	0.0093	<0.000043[c]
Ratio	1.8	470	>1000
Ligand type	Antagonist	Partial agonist	Full agonist

Fig. 2. *Oxadiazole-based muscarinic ligands*

[a] Dissociation constant (K_{app}, μM) for the displacement of [^{3}H]NMS from rat cortical membranes. [b] Dissociation constant for the displacement of [^{3}H]OXO-M. [c] Radioligand depletion in the binding assay causes an underestimate of true affinity.

binding, but significantly more so at the low-affinity state leading to partial agonist behaviour. Finally, introducing an amino substituent into the oxadiazole ring, and changing the quinuclidine framework to the less lipophilic azanorbornane system, affords the most potent and efficacious muscarinic agonist known, primarily by enhancement of binding to the high-affinity state of the receptor. The figure of 0.043 nM is an underestimate of the true affinity of amino-oxadiazole for the OXO-M-labelled state, because of ligand depletion in the binding assay. When the determination was repeated using lower protein concentrations, the affinity was apparently increased and this gives a binding ratio in excess of 1000 [1]. That this compound behaves as a full agonist was confirmed by the magnitude of its PtdIns response, which was measured to be 150–170% to that seen with acetylcholine and the effect was completely atropine sensitive [1]. This latter compound is an excellent agonist ligand to use in docking experiments, not only because of its exquisite activity, but also because, unlike the natural neurotransmitters in general, this molecule has only one rotatable bond, so that the geometry of the agonist pharmacophore is well defined. For reasons that will become obvious later, the presence of at least one rotatable bond is seen as highly desirable to allow the necessary molecular tumbling associated with agonist behaviour.

As has been obvious for some time, all muscarinic ligands require the presence of a cationic head group which we now know interacts with an Asp carboxyl group on the receptor and this is seen as the primary interaction. Until these compounds were discovered, it was generally believed that only quaternary ammonium muscarinic agonists had the ability to induce a full response in the cortex. The quinuclidine ring and related azabicyclic systems were chosen since quantum mechanical calculations showed us that, in their protonated form, the charge distribution most closely resembled the spherically symmetrical distribution seen in quaternary methylammonium species. Sites for secondary interactions, and those which allow us to distinguish between agonists and antagonists, are provided for by the oxadiazole moiety. Progressive removal of the heteroatoms from the ring, by making oxazoles and furans, for example, caused dramatic reduction in predicted efficacy by loss of binding to the high-affinity state of the receptor, while low-affinity-state binding was relatively unaffected [1, 10]. It was concluded that two H-bond acceptor sites on the ligand were necessary for full agonist behaviour and, furthermore, these sites were localized in the vicinity of the two ring nitrogens. The role of the oxadiazole ring substituent is simply to influence the H-bonding capability of the ring nitrogens. Electron donating groups, such as amino groups, potentiate this effect. More precisely, it was shown that efficacy at the MR correlated directly with the negative electrostatic potential adjacent to these two atoms. Thus two H-bonding interactions are necessary for binding with high affinity to the agonist-binding site with little or no dependency on lipophilic binding (Fig. 3). Either or both of these interactions are relatively unimportant for binding to the low-affinity state of the receptor, that is, for antagonist binding. Instead, increased reliance upon lipophilic binding in the vicinity of the acetylmethyl group of the endogenous ligand characterizes antagonist behaviour with partial agonists falling between these two extremes.

In fact, we can give a more generalized ligand-binding map for any monoamine neutrotransmitter or antagonist interacting with its specific receptor (Fig. 4). We have already discussed the muscarinic receptor where this lipophilic domain is simply CH_2CH_2 in the natural agonist. In the case of the β-adrenergic receptor, this is replaced by a planar aromatic group, the cationic head group is a protonated amine and the catechol provides two H-bonding interactions, one an acceptor and one a donor.

This has recently been supported by additional mutagenesis studies on the β-receptor by Strader and colleagues [2]. Several single point-mutations were made replacing potential H-bonding side-chains with Ala to give three groups of mutants: the first failed to produce immunoreactive protein, the second

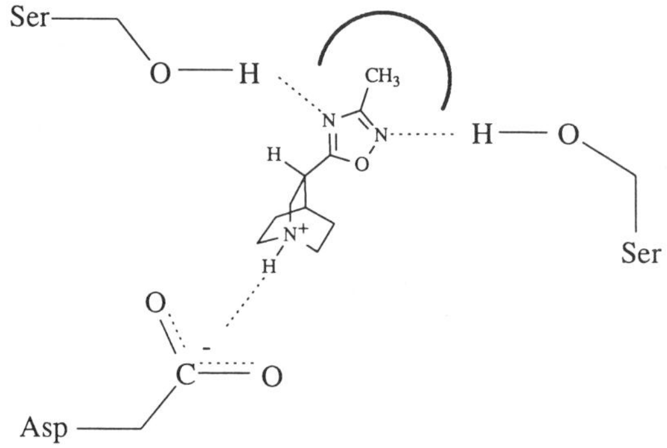

Fig. 3. *Hypothetical binding site for muscarinic ligands involving the Coulombic primary interaction between cationic head group of ligand and Asp residue of protein*

Secondary H-bonding sites are represented by two putative Ser residues. The lipophilic domain on the receptor required for tight-binding antagonists is also shown adjacent to the oxadiazole methyl substituent.

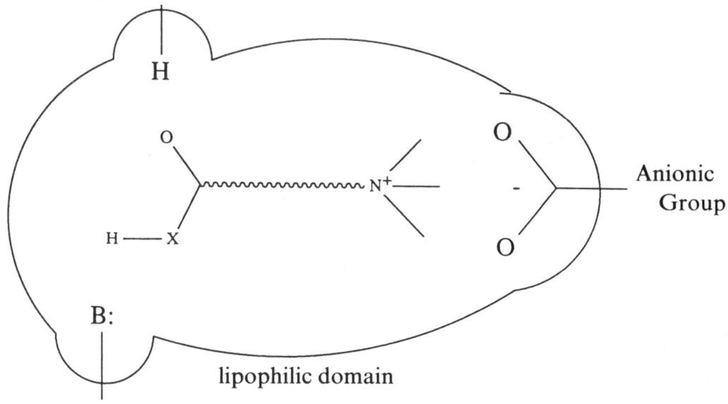

Fig. 4. *Generalized agonist binding map for monoamine neutrotransmitter ligands*

showed no change in ligand-binding characteristics and the third caused changes to both the affinity and efficacy of the bound ligand. The most interesting of the latter category were two Ser residues on helix V, Ser-204 and Ser-207, which appear to be specific for receptors which bind catechol ligands. Only agonist and not antagonist affinity was reduced in the mutant receptors, which is presumably a reflection of disruption of a H-bond between ligand and mutant receptor. More interestingly, however, was the fact that efficacy was also reduced so that interaction at these two secondary sites by the ligand is obligatory for a fully coupled receptor. One might speculate that simultaneous mutation of both Ser-204 and Ser-207 would lead to an uncoupled receptor.

In support of a role for H-bonding interactions are some binding data, also by Strader, for β-ligands which have been structurally altered to remove potential H-bonding sites. Removal of both hydroxyl groups of the catechol moiety gave a ligand which displayed almost identical affinity for all three proteins (wild-type, Ala-204 and Ala-207 mutants), but with marked reduction in efficacy. Efficacy was restored by adding back one or other of the phenyl-ring hydroxyls. It was also evident from the data that the *meta*-hydroxyl interacts with Ser-204 while the *para*-hydroxyl pairs with Ser-207.

In summary, it is possible to differentiate between ligands of varying efficacy as follows. The cationic head group is an essential element for all ligands. Full agonists utilize H-bonding interactions, usually two, whose nature and topology varies depending upon the receptor type. Agonists are small hydrophilic molecules. Antagonists rely on lipophilic binding energy to stabilize binding to the receptor and are therefore usually large molecules with some region of high bulk lipophilicity. This would also imply that the design of antagonists from agonists, or vice versa, should be straightforward and, indeed, that ligands for one receptor may be used to design ligands for a second receptor.

Ligand-Docking Experiments

The model of the MR described earlier has been used to test some of the ideas generated above regarding the nature of secondary binding interactions.

The three ligands of widely differing efficacy have been docked into the receptor model to rationalize why such molecules should have such a diverse effect on receptor activation. Model building for the three ligands was based on X-ray diffraction data. In each case, molecular mechanics calculations revealed that the barrier to rotation about this single bond is less than 1 kcal mol^{-1}, so that this bond was allowed to relax during the docking procedure.

First, it is proposed that an antagonist may bind to the receptor in a manner referred to as Mode 0 (Fig. 5*a*). For reasons that will become apparent, the antagonist molecule, An, has been deliberately drawn bound above Asp-105. The mechanics of docking were as follows: for simplicity and consistency, each ligand, in this case the antagonist, was docked with the relevant Asp residue in a H-bonding fashion involving the NH^+ of the protonated ligand and the Asp carboxyl O atom and the N–O distance kept constant. At this stage, the Asp side-chain dihedrals, the rotatable bonds of the ligand and

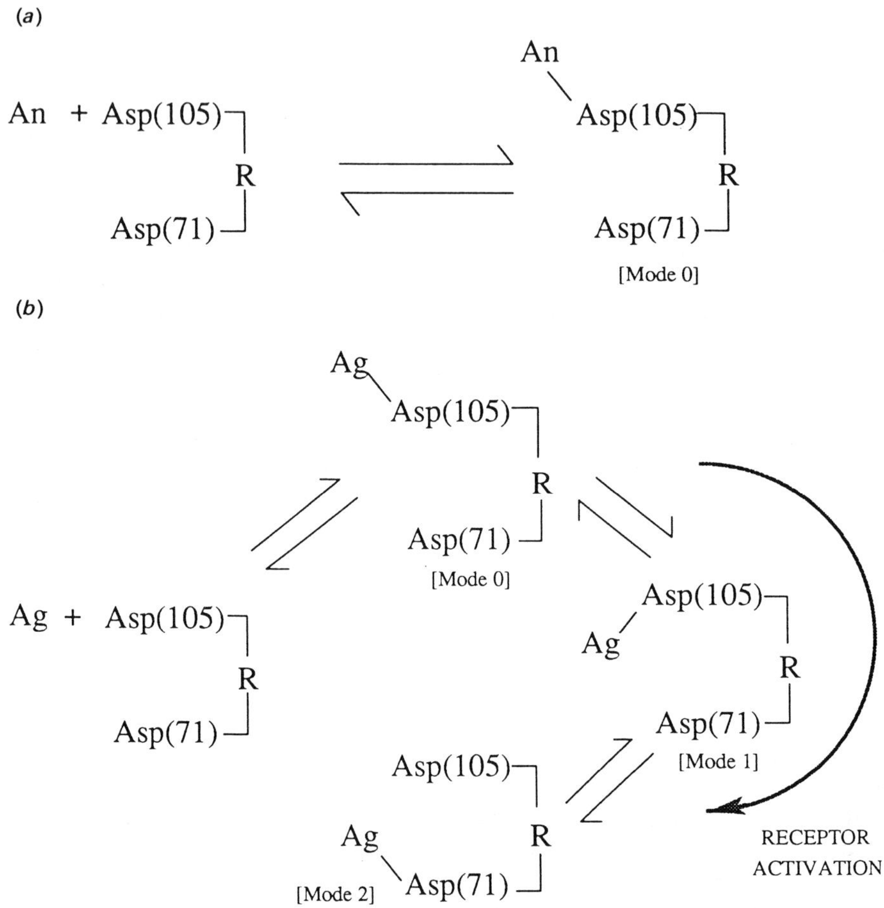

Fig. 5. *Proposed sequence of equilibria that underscore receptor activation following agonist binding*
At some level beyond the initial ligand–receptor complex, conformational changes in the protein induced by ligand equilibration putatively triggers receptor coupling to G-protein. (*a*) Shows the final result of antagonist (An) binding. (*b*) Agonist (Ag) binding.

the side-chains of adjoining residues, were included in a Van der Waals minimization procedure to give a docked ligand free from bad contacts. The extended lipophile present in all tight-binding muscarinic antagonists can be manoeuvred into potential lipophilic pockets close to the receptor surface and involving primarily helices I, II, III, VI and VII. One such pocket is provided for by a combination of a Phe and two Trp groups, and may well explain the preference for aromatic lipophiles, such as are present in scopolamine and QNB, and thus the relatively lower affinity for antagonists devoid of such groups, such as dicyclomine. For antagonists, this is the end of the story— they effectively plug the entrance to the receptor and are unable to participate in any conformational changes that lead ultimately to receptor activities.

It is anticipated that agonists also recognize the low-affinity state of the receptor in the initial collision complex; i.e. they also bind in Mode 0, but with comparative lack of bulk at the entrance to the receptor. At some stage, however, the conformational lability of the side-chains of the functional Asp-105 (in MR) will drag the relatively small molecule into a new position with the cationic head group some 12 Å away from the surface of the protein. This is termed Mode 1 binding (Fig. 5*b*) and represents the first agonist-specific binding mode. Stabilization of binding at this point can no longer rely upon lipophilic binding, since the oxadiazole and its substituent find themselves in a highly hydrophilic climate provided by the side-chains of Ser, Thr and Asn residues which seem to litter this locus of the receptor. Not coincidentally, the H-bond acceptor N atoms of the oxadiazole ring fall roughly mid-way between Asp-105 and Asp-71. This then is Mode 1 binding and is probably the state most heavily occupied by ligands with intermediate efficacy, i.e. partial agonists. With the exception of just one Ser, these very same residues are able to stabilize the equilibrium between binding with Asp-105 and, for agonists, Asp-71, which is assumed to represent the high-affinity state of the receptor, here called Mode 2. It is physically impossible for the bulky antagonist molecule to achieve such a binding mode. At some stage during the transition from Mode 0 to Mode 2, it is proposed that the backbone undergoes a gross change in conformation, possibly triggered from those regions where helical structure is weakest close to the proline residues, i.e. synchronous with receptor activation.

Ligand-Induced Receptor Activation

This sequence of events may be summarized as follows. Any ligand, irrespective of its efficacy binds in Mode 0 at Asp-105 as shown in the first equilibrium process. Only if the ligand is an agonist, as shown here, is the second equilibrium achieved (Fig. 5*b*) with the ligand moving down into the receptor-binding channel. Within this highly hydrophilic region, H-bonding interactions are used to assist in the shuttling process by which the agonist becomes attached to the lower Asp, Asp-71. Occupation of a position between these two functional Asps is assumed to lead to coupling to G-protein.

This has three consequences. First it requires an alternative explanation of the NMS/OXO-M binding assay to that originally proposed [8], since it implied that antagonists showed equal affinity for both states of the receptor leading to a binding ratio close to 1. What this model is suggesting is that antagonists cannot possibly displace agonist binding from the high-affinity state of the receptor directly. Instead, antagonists displace radiolabelled agonist indirectly by continually disturbing the equilibria in favour of low-affinity binding (Mode 0) and thus displace both NMS and OXO-M from the same affinity state of the receptor. Secondly, it provides a qualitative explanation of partial agonists whose efficacy will depend on the exact position of the third equilibrium. For full agonists, this equilibrium will lie almost entirely in favour of Mode 2. Finally, since achievement of Mode 2 requires mutual conformational shifts in both agonist and receptor, it is unlikely that completely rigid analogues of acetylcholine (for the MR) such as AF-30, cannot function as high-efficacy agonists in tissues such as cerebral cortex where there is limited receptor reserve.

I thank my ex-colleagues in Medicinal Chemistry at Merck (R. Baker, A. MacLeod, K. Merchant, G. A. Showell, R. Snow and L. Street) for superb collaboration during this work. I am indebted to S. Freedman for his biochemical insight and G. Smith (Merck, West Point) for generating the original receptor model.

References

1. Saunders, J. & Freedman, S. B. (1989) *Trends Pharmacol. Sci.* (*Suppl.*) (Dec.), 70–75
2. Strader, C. D., Sigal, I. S. & Dixon, R. A. F. (1989) *Trends Pharmacol. Sci.* (*Suppl.*) (Dec), **10**, 26–30
3. Dixon, R. A. F., Strader, C. D. & Sigal, I. S. (1987) *Ann. Rep. Med. Chem.* **23**, 221–233
4. Henderson, R. & Unwin, P. N. (1975) *Nature* (*London*) **257**, 28–32
5. Findlay, J. B. C. & Pappin, D. J. C. (1986) *Biochem. J.* **238**, 625–642
6. Kubo, T., Bujo, H., Akiba, I., Nakai, J., Mishina, M. & Numa, S. (1988) *FEBS Lett.* **241**, 119–123
7. Fraser, C. M., Wang, C. D., Robinson, D. A., Gocayne, J. D. & Venter, J. C. (1989) *Mol. Pharmacol.* **36**, 840–847
8. Bonner, T. L., Young, A. C., Brann, M. R. & Buckley, N. J. (1988) *Neuron* **1**, 403–410
9. Freedman, S. F., Harley, E. A. & Iversen, L. L. (1988) *Br. J. Pharmacol.* **93**, 437–445
10. Saunders, J., Cassidy, M., Freedman, S. F., Harley, E. A., Iversen, L. L., Kneen, C., MacLeod, A. M., Merchant, K. J., Snow, R. J. & Baker, R. (1990) *J. Med. Chem.* **33**, 1128–1138

Biochem. Soc. Symp. **57**, 91–97
Printed in Great Britain

The Expert System Approach to Predicting Protein Structure

B. ROBSON*, J. BALL, R. V. FISHLEIGH, P. GREANEY, J. LI, A. MARSDEN, E. PLATT and J. L. POOL

*Proteus Molecular Design Limited, 48 Stockport Road, Marple, Cheshire SK6 6AB, U.K. and *Proteus International plc, 48 Stockport Road, Marple, Cheshire SK6 6AB, U.K., Department of Biochemistry and Molecular Biology, University of Manchester and The Technical University of Denmark*

Synopsis

Prediction of protein structure is an open-ended problem. Since an approach from first principles cannot be taken in reasonable computer time, short-cuts using further data are necessary. Such data include information about the specific protein in question, and information in databases which are about proteins in general. Is it possible to write a general, flexible 'superalgorithm' which would suit most circumstances? If so, it would seem likely to overcome one of the most understated but nonetheless greatest difficulties associated with molecular modelling and computer-aided drug design — reproducibility. To this end, a 'polymorphic programming environment' has been developed which represents both an expert system and a high-level language for theoretical chemists and molecular biologists. This language is GLOBAL (Ball *et al.*, 1990). In a series of earlier studies, and more recently by means of GLOBAL itself, the nature of reproducibility and its rather surprising limits have been explored, and in general the current status and future potential of protein modelling have been examined.

Introduction

This paper considers the general question of whether complex projects in drug design, protein engineering and other theoretical chemistry disciplines can be fully automated, replacing the need for human experts and requiring only occasional upgrading by a limited number of high-ranking experts. The specific examples given are confined to protein modelling. The question of the quality of results remains the outstanding one, but it is primarily the question of automation which is considered here, a matter which is of considerable scientific and technological importance. In practice, complex molecular modelling projects, involving a variety of programs and expert interpretation of output, are rarely if ever fully reproducible because of the interactive element. In particular, an expert using molecular graphics to deduce how a ligand docks to a protein-binding site will be unlikely to reproduce his docking procedure, and probably even the details of the result, on two consecutive days. In

* To whom correspondence should be addressed.

our view, even the work described in detailed papers on molecular modelling topics by distinguished authors are as difficult to duplicate quantitatively as experimental work, and in some cases more so. Though studies are carried out and reported in good faith, it cannot be ignored that this in practice breaks one of the most fundamental scientific principles: the need for reproducibility.

Theory and Method

Several approaches have been taken to explore the reproducibility and the possibility of automating molecular modelling projects. In the early approaches, we simply aimed at reproducing quantitatively the results of scientific publications. The results of some authors are more reproducible than others, partly because of differences in the level of detail given, in what the reader is expected to assume, and in the quality with which detailed data tables are proof-read. In our personal experience, the papers of the Flory school are particularly easy to reproduce and were concerned with calculating physicochemical properties from interatomic potential functions. This made them very suitable subjects for this kind of quantitative analysis. In an interesting early study with Dr A. T. Hagler we were indeed able to reproduce quoted calculated properties, starting from the parameters for molecular geometry and interatomic forces as given; however, it was only by reproducing every step of the calculation that we discovered some 'down-stream' errors that would have fooled readers commencing the calculation at an intermediary stage. For example, we noted that two rather similar looking potential surfaces (energies expressed as a function of conformation) in a paper by Brant *et al.* (1967) had been accidentally inverted with respect to the figure legends, in a 'typographical' error. The inclusion of the consideration of such errors in a deeper analysis of reproducibility is nonetheless essential. We also noted that the quality of the results given in that paper depended sensitively on the resolution of thermodynamic sampling as actually used to evaluate conformation-dependent properties and, as it happened, the agreement with experiment got worse with improved sampling. We concluded that the authors had made an arbitrary choice in a step of the procedure to which the quality of the results were at least slightly sensitive.

To learn how to minimize irreproducibility in such research work Proteus sought first to obtain a mathematical model of reproducibility. Any theoretical modelling or prediction research project was considered only in terms of a written set of instructions for it, such instructions constituting a protocol—a kind of 'superalgorithm'. The relations between (i) the structures of a protocol and (ii) the derivatives of quantitative results with respect to various kinds of perturbation of each protocol step were explored. The general qualitative findings were as expected: some steps in any protocol ultimately lead to results rather insensitive to the precise choice of values used, or interpretation of a previous step, but some steps are 'hot spots' in which the smallest perturbation of procedure or values can lead to radically different results. Sometimes this crucial character of a single step depends on the overall choice of procedure. For example, the choice of method for secondary structure prediction could lead to quite different three-dimensional structures in an overall protein structure prediction product, if appropriate care is not used in designing the overall modelling protocol.

A deeper analysis is far less trivial, however, and requires the mathematical tools and insights developed by W. R. Ashby and others, namely systems theory and 'superstability theory'. Ashby (1952) studied mechanical and electronic models; to develop this more detailed insight into protocols also demands automation. Largely to automate complex research projects in pharmacology and molecular biology as a desirable goal in itself, and concomitantly to explore and minimize the irreproducibility inherent in protocols, an unambiguous protocol language has been developed. It consists of an expert system and high-level language, and is also a 'polymorphic programming environment', in that it also includes an operating system and other facilities, here specifically tailored for molecular work. This is both a tool for carrying out projects in the case of clearly defined problems, and also a development environment in which more complex, open-ended problems can be tackled. The language used for its execution and development is GLOBAL (Ball *et al.*, 1990) and is capable of interactive as well as batch use. The aspect which makes it specifically a tool for theoretical chemistry and molecular modelling, rather than simply a very high-level language, is that operations on molecular models, data and simulations are effectively at the same level as normal computing operations. Hence `$X = 2.5*COS(THETA);` `$SCRAP LINE 26;` and `$FOLD "A.V.L.L.G.C.S.T.";` are all basic instructions at the same apparent level. The need at least in the most fundamental form of GLOBAL, for the symbols '$', and ';' is to do with its functional programming system (FPS) capabilities (see below). The language is fundamentally redefinable and a unique 'artificial intelligence' language feature is the ability to replace the above basic or 'canonical' instructions by input in human language which is parsed and interpreted, or mixed freely with it. This sophistication is not accidental: it is essential that a complex expert system be accessed in a user-friendly way by the inexpert. More importantly, it turns out to be important in mapping human expertise to protocols, and maps more directly to the way a human expert will describe the protocols, i.e. in human speech. In brief, if human language cannot be parsed, interpreted and mapped to unambiguous protocols written in canonical GLOBAL, then the problem is not yet well defined. However, use of 'fuzzy logic', FPSs with self-editing and recursion, and neural network capabilities, does allow a middle ground in cases of uncertainty and ambiguity by mimicking in a simple way some features of human brain function.

One of the most difficult aspects of developing an automated approach in theoretical chemistry, pharmacology and molecular biology is simply the enormous variety of types of information and data which must be communicated. The outstanding difficulty of processing and passing these data which is encountered in molecular design work was overcome by including an FPS capability in the language. `$FOLD "A.V.L.L.G.C.S.T.";` might be replaced by:

```
$FOLD $SEQUENCE "BROOKHAVEN " "2INS" "CHAIN" "A" ;;
```

or by:

```
$MY_SEQUENCE = $SEQUENCE "BROOKHAVEN" "2INS"
        "CHAIN" "A" ;;$FOLD MY_SEQUENCE;
```

to fold the insulin A chain. Passing of data up through progressive embedded levels can involve quite complex operations relieving the user of the responsibility to process the data. At a still further level, the command:

```
FOLD CHAIN A OF 2INS
```

can be easily defined in terms of general arguments to relieve the need of the user even to understand the GLOBAL programming principles. The GLOBAL language is used to drive and control the BioEngine, which is a molecular modelling system consisting of both hardware and software.

Results

Several successes have been achieved in maximizing the automation of complex drug-design projects. The protocols written in the GLOBAL language are quite long and complex in such cases, even though care has been taken in developing a user-friendly GLOBAL which makes them easy to read and understand. Here we seek to consider examples which can be written in some 500 lines of GLOBAL or less. At this symposium and elsewhere, Chris Sander stated verbally that 'the problem of protein structure determination by homology is virtually solved' (Sander, 1990). If this is the case, as seems reasonable, then it argues strongly for the possibility of a general, flexible super-algorithm which is not too complex and difficult to analyse as to stability. We confirm that in the case of a protein where a homologue exists, and the structure for that homologue has been defined, the determination of the structure of the protein in question is largely a matter of following a series of defined steps. These protocols can indeed be developed into a number of procedures, which can be fully automated. There will nonetheless be decision-taking steps in these procedures which will involve expert system principles. For such reasons (not unique to protein modelling by any means), GLOBAL is also an 'expert system' capable of handling fuzzy and quantitative logic, though the degree of success of this, and certain important implications for modelling, will be described elsewhere.

Our earlier approaches involved the Bayesian subjectivist method popularized by the Department of Mathematics in Cambridge, Dr Simon French at Oxford and B. Robson at Manchester (Robson, 1974). The latter was the first application as a predictive method in the molecular sciences. Though these lead in turn to a first theory of reproducibility and irreproducibility, our studies were eventually overtaken by the popular development of a more general theory of non-linear mathematics: chaos. The most important and general findings of our work are simply that even relatively simple protocols can be subject to the 'laws of chaos' and problems inherent in any feature displaying elements of feedback or non-linearity in general. That is, results can occasionally vary when the same algorithm is slightly perturbed or when executed on different machines, because of certain critical steps sensitive to machine precision in number representation and manipulation. At one time, such sensitivity would be considered simply as bad programming, but the problems are deeper than simply sensitivity to number. Bad programming is too trite an answer and the problem is more fundamental: indeed, even algebraic mathematics can be shown to have a 'chaotic' aspect (Chaitin, 1990). One can of course choose algorithms which damp the consequences of this sensitivity, providing one accepts reproducibility at a slightly coarser level. Fortunately, in molecular modelling and prediction work this is within a scientifically adequate level of detail, and typically well within the errors encountered in X-ray crystallography. This was explored in a different way in protein modelling by choosing three different approaches in the modelling of thrombin by homology (two approaches were, to different extents, particularly interactive). Rather similar results were obtained, though not at a finite level of detail (Robson, 1980; and E. Platt, B. Robson, R. V. Fishleigh, P. Greaney & A. Marsden, unpublished work).

A similar early study of the modelling of human α-lactalbumin by homology with hen egg white lysozyme (Robson & Platt, 1987), using a system predating the Bioengine (Robson & Platt, 1986; Robson *et al.*, 1987), but still essentially an automatic procedure from the alignment stage, gave a 2.8 Å r.m.s. deviation (using the crystallographers' method) for all atoms between one of the models and the X-ray co-ordinates of baboon α-lactalbumin (Robson & Platt, 1990). It should be noted that this modelling study was performed and the results were independently filed before the X-ray results became available. From a general 'systems theory' viewpoint it is of interest that the operation of 'energy refining' the X-ray structure, as involving an energy relaxation combined with r.m.s. fit to initial X-ray coordinates, is a 'simple catastrophe'. The number of multiple minima are greatly increased in the function surface, and the main and obvious consequence is that there are two principal best solutions: the energy is bad and the fit good, or vice versa. The difficulty only disappears in the ideal of a molecular model with perfect bond geometry and interatomic potential functions, and perfect initial data. This is neither an achievable situation nor an interesting one.

Here we are mainly concerned with some specific points of importance to biochemists and molecular biochemists. Some more biological aspects were harder to automate than others, simply because there is no consensus agreement between experts about the best approach. What is generally recognized is that the problem of recognizing relations and determining structure where only parts of proteins are homologous, or even where no protein is homologous even in part, might be overcome by considering that very few types of structural domain may exist in nature. The precise method for exploiting this is, however, still a matter of opinion. In cases where parts of proteins are obviously homologous these are often parts common to many proteins with a particular function, e.g. DNA binding (where the structure might be a Rosman fold, helix bundle, zinc finger, leucine zipper, etc.), ATP binding, etc. Identifying these and other domains in the case of weak, ambiguous or 'cryptic' sequence homology is complex, involving more decision-taking steps and more possibilities at each step. A very detailed analysis of the problem has been performed by us in Bayesian terms, and will be reported elsewhere. In brief, the notion of statistical significance in sequence relations is not an absolute, but can be shown to be dependent on the purpose to which the results are to be put, i.e. on the nature and purpose of the protocol overall. Even when a number of possible domains can be identified, however, the extent to which they are distinct or similar, and the extent to which they can be 'cannibalized' or adapted to form the basis of the new model, is a difficult problem. Such difficulties in domain recognition and use can be overcome by using a 'theorem proof' approach. Again, in brief, the formulae which Robson & Garnier (1986) developed for describing supersecondary and domain structure are interpreted as theorems, which must be proven one from the other by valid steps. By analogy with a common puzzle for children, if the word Jane can be converted to Mary by changing one letter at a time, each resulting in a valid word, then Mary is said to be provable from Jane and Mary and Jane are said to be equivalent 'in essence'. Valid operations for transformation have been identified, mostly by analysis of protein evolution, though often remarkably consistent with stereochemical reasoning and considerations of chain orientation and continuity. This approach promises to be a major area for sophisticated automation of modelling proteins without obvious homology.

As mentioned above, the problem of protein structure determination by homology is virtually solved. This is the case where a strong homology can be detected along most of the protein length. The cases where weaker or more fragmentary homologies are detected, or homology exists for only a part of the sequence, is more complex to solve.

The initial stages of any modelling study involve analysis and processing of external data. This includes identification of homology, in terms of sequence and conformation, by alignment. Conformational homology scans in particular can identify conserved regions of the sequence, and indicate non-conserved regions, deletions and insertions. If the relationship with other proteins is weak, then more sensitive homology scans are required to identify domains and subdomains which may be of use in the initial stages of modelling. If a strong relationship is established, this can be used as a bias for the initial modelling stages.

The simplest example of 'cannibalizing' or extensively adapting a domain is probably in regard to serine proteinase barrels, well explored in our modelling studies over many years. We have been particularly interested in modelling thrombin and, although this belongs to the family of serine proteinases which have clear sequence and structural homology, some of the basic principles of adapting domain motifs are the same. Having identified the conserved regions and the regions of homology, the stages to follow are: (*a*) to fit the whole chain of thrombin to the chosen homologue, allowing for insertion/deletions; (*b*) to substitute the correct side-chains for thrombin and carry out a r.m.s. fit of the backbone C_α, replacing non-conserved residues and insertion/deletions regions, and to allow relaxation of the structure to minimization; (*c*) to represent the side-chains in full rather than as one atom, and to assign side-chain angles, and to carry out refinement using target functions; and (*d*) refinement using molecular dynamics and similar simulations. Of course, the problem of protein modelling where homology is weak or fragmentary is less clearly defined. In this case, having identified homologies where they exist, possibly to sections of several different proteins, the following stages could guide the modelling: (i) r.m.s. fitting of the fragments where homology has been identified; (ii) fragment assembly and representation of non-homologous regions; (iii) adoption of full side-chain representation and choice of side-chain starting angles; and (iv) relaxation and refinement using various techniques including molecular dynamics. There are many more decision-taking steps, and many more variables which could influence the choice made at these steps. It is perhaps to these areas that modelling experts could devote their attention in future. The primary conclusion to be sought by that study was, however, whether the above procedure as applied to thrombin could be carried out as a wholly automatic process. This was indeed so, and it could be done as a fairly general protocol by no means specific to serine proteinases. The command:

PLEASE FOLD THROMBIN

was all that was required and as yet we have no reason to doubt that:

PLEASE FOLD HIV PROTEASE

would not fare just as well, using the same definitions of the verb FOLD as a generalized set of operations written in GLOBAL. This and other studies are in progress, particularly in cases, as for thrombin itself, where the modelling precedes the X-ray result. In such cases, as with thrombin, results are independently analysed and assessed and conclusions will be reproduced in due course.

References

Ashby, W. R. (1952) *Design for a Brain*, Wiley, New York

Ball, J., Fishleigh, R. V., Greaney, P., Li, J., Marsden, A., Platt, E. & Robson, B. (1990) in *Chemical Information Systems* (*Beyond the Structure Diagram*) (Bawden, D. & Mitchell, E. M., eds.), Ellis Horwood Press

Brant, D. A., Miller, W. G. & Flory, P. J. (1967) *J. Mol. Biol.* **23**, 47–65

Chaitin, G. (1990) *New Scientist* March 24, **125**, 44–46

Robson, B. (1974) *Biochem. J.* **141**, 853–867

Robson, B. (1980) *Trends Biochem. Sci.* **4**, 240–244

Robson, B., Platt, E., Fishleigh, R. V., Marsden, A. & Millard, P. (1987) *J. Mol. Graphics* **5**, 8–17

Robson, B. & Garnier, J. (1986) *Introduction to Proteins and Protein Engineering*, pp. 91–116, Elsevier Press, Amsterdam

Robson, B. & Platt, E. (1986) *J. Mol. Biol.* **188**, 259–281

Robson, B. & Platt, E. (1987) *J. Comput. Aided Mol. Des.* **1**, 17–22

Robson, B. & Platt, E. (1990) *J. Comput. Aided Mol. Des.* in the press

Sander, C. (1990) *Biochem. Soc. Symp.* **57**, 25–33

Biochem. Soc. Symp. **57**, 99–121
Printed in Great Britain

Basic Design Features of the Parallel αβ Barrel, a Ubiquitous Protein-Folding Motif

SHOSHANA J. WODAK*, IGNACE LASTERS†, FRÉDÉRIC PIO*‡ and MICHEL CLAESSENS†§

* *Unité de Conformation de Macromolécules Biologiques, CP160, P2, Université Libre de Bruxelles, Ave. P. Héger, B-1050 Brussels, Belgium and † Plant Genetic Systems, CP160, P2, Université Libre de Bruxelles, Ave. P. Héger, B-1050 Brussels, Belgium*

Synopsis

Basic design features of the β-sheet portion in parallel αβ barrels in known protein structures are analysed in the context of a model of a regular hyperboloid. A formal description of the relationships between β-sheet twist, number of strands in the sheet and barrel dimensions is derived, and the underlying physical principles are rationalized.

Results suggest that the major constraints on the geometry of the β-sheet portion of the barrel come from the requirements to have optimal H-bonding interactions between β-strands and to closely pack amino acid side-chains in the barrel interior so as to exclude bulk water. In addition, we show how the hyperboloid model and the ensuing formalism can serve to derive useful geometric and graphic tools for computer-aided protein design *de novo*. We then illustrate how these tools are used to determine that the requirement to have a closed regular eight-stranded β-sheet surface imposes no particular constraints on the geometry (φ, ψ angles) of the polypeptide backbone. Understanding the role of the amino acid sequence in determining the observed structures remains a major challenge. Detailed comparisons of known αβ-barrel structures (and amino acid sequence) with each other, and with polypeptide fragments from other protein crystal structures, reveal only a limited number of common sequence–structure motifs. These belong to characteristic αβ1 and αβ3 loop families previously described in αβ proteins, and occur at least once in nearly all the αβ-barrel structures examined.

Introduction

Analysis of known protein structures [1–3] has shown that they are often organized into smaller units called domains, and that these domains display a limited repertoire of folding motifs. One such motif is the eight-stranded parallel αβ-barrel, found to occur in proteins of the α/β class. First observed in

‡ Present address: University of California, Department of Medicine, Division of Endocrinology and Metabolism, San Diego, La Jolla, CA 92093, U.S.A.

§ Present address: COMETT Technical Office, avenue de Cortenbergh, 71, B-1040 Brussels, Belgium.

triose phosphate isomerase [4] and therefore termed the 'TIM barrel', it has since been found in at least 15 more proteins and this number is increasing steadily.

In addition to TIM, the eight-stranded $\alpha\beta$ barrel occurs in one of the three domains of the subunit in tetrameric pyruvate kinase [5] and in dimeric trimethylamine-dehydrogenase [6]. It also occurs in aldolase [7], taka-amylase [8], α-amylase [9], glycolate oxidase [10], ribulose-1,5-bisphosphate carboxylase (Rubisco) [11], flavocytochrome-b_2 [12] and xylose isomerases [13–15], as well as in cyclodextrin glycosyl transferase [16], tryptophan synthase A subunit [17], mandelate racemase and mandelate dehydrogenase (D. J. Neidhart, G. L. Kenyon, J. A. Gert & G. A. Petsko, unpublished work). The bifunctional enzyme *N*-(5′-phosphoribosyl)-anthranilate isomerase-indoleglycerol-phosphate synthase has two $\alpha\beta$ barrels with different catalytic activities in the same monomer unit [19]. An incomplete version of the barrel is found in two other enzymes. In muconate cycloisomerase there are only seven helices in the barrel instead of the usual eight [20], and in the cellobiohydrolase II from *Trichoderma reesei*, one β-strand in the barrel is missing [21]. It is still somewhat unclear whether the enzyme enolase contains a classical $(\alpha\beta)_8$-barrel motif, as initially thought [22], or a variant with the topology $\beta\beta\alpha\alpha(\alpha\beta)_6$, where the second β-strand and the first helix run antiparallel to all other β-strands and to the α-helices, respectively [23].

Among the 16 or so classical $\alpha\beta$ barrel motifs some display homology in amino acid sequence and/or in function. But in general, sequence homology among the motifs is very low and they perform different enzymic reactions on a variety of substrates. This may be taken to mean that the parallel eightfold $\alpha\beta$-barrel is a particularly robust fold on to which different functionalities can be readily designed. But how has this fold arisen? Has it evolved from a common ancestor [24] or through convergent evolution, or, alternatively, through a process of cut and paste of structural and/or functional subdomains [25]? Experimental approaches, which probe into this question have been recently reported. They include the synthesis of circularly permuted homologues of *Escherichia coli* *N*-(5′-phosphoribosyl)-anthranilate isomerase, which folds as a separate monomer in this organism [26,27], and the synthesis *de novo* of an artificial polypeptide made up of repeats of a putative β/α-forming amino acid sequence [28].

Theoretical analyses of the structural and energetic requirements for folding of $\alpha\beta$ barrels should be a valuable complement to these approaches. Indeed, they offer a framework for rationalizing experimental observations and for designing experiments that test the theory. Here we present a synthesis of such theoretical analyses with emphasis on work performed in our own laboratories.

In the first analysis, the basic design features of the β-sheet portion in parallel $\alpha\beta$ barrels from nine known protein structures are studied in the context of a model of a regular hyperboloid which previously has been shown to yield a good description of the TIM barrel [29]. A formal description of the relationships between β-sheet twist, number of strands in the sheet and barrel dimensions is derived, and the physical constraints that govern these parameters are rationalized [30]. It is furthermore shown that the hyperboloid model can be used as a scaffold to construct idealized parallel β-barrels [31]. To facilitate the design *de novo* of such structures, a collection of modelling tools has been developed allowing us to map secondary structure elements on to the scaffold surface and to perform rotation and translation operations about user-defined axes, while evaluating their contribution to the conformational energy of the system. These tools are then applied to analyse the ϕ, ψ requirements for designing a symmetric eight-stranded parallel β-barrel with optimal H-bonding between adjacent β-strands.

Understanding the role of the amino acid sequence in determining the observed $\alpha\beta$ barrel structures remains a major challenge. The wide sequence diversity compatible with this ubiquitous fold suggests that it may be a rather non-specific and stable motif that is quite tolerant to sequence variations. To understand the limits of this tolerance then becomes the central question.

As a first step towards investigating this question, an analysis aimed at finding local structure and sequence motifs in an $\alpha\beta$ barrel structure has been initiated. The approach, which is based on comparing fragments from $\alpha\beta$ barrels with fragments in a large number of protein structures, reveals so far only a limited number of recurrent sequence patterns associated with specific local structure motifs. These belong to characteristic $\alpha\beta1$ and $\alpha\beta3$ loop families previously described in $\alpha\beta$ proteins [32], and occur at least once in nearly all the $\alpha\beta$-barrel structures examined so far.

Basic Design Features of the Eight-Stranded Parallel β-Sheet in αβ-Barrels

Geometrical properties of observed parallel β-barrel

The β-sheet portion of the parallel $\alpha\beta$ barrel constitutes itself a closed barrel-shaped surface formed by eight parallel β-strands. To analyse the geometrical properties of this surface, regular hyperboloids, defined by the quadratic equation:

$$(x/a)^2 + (y/b)^2 - (z/c)^2 = 1 \qquad (1)$$

where a and b are the major and minor semiaxes and c the curvature in the direction of the hyperboloid axis, have been fitted to C_α atoms of the β-strands composing the sheets in parallel $\alpha\beta$ barrels of nine proteins structures. The fitting procedure which has been previously described [30] involves computing least-squares lines passing through individual β-strands, and then using these lines to derive the relevant parameters that define the hyperboloid.

Fig. 1 illustrates the fit between β-strand axes and the generating lines of the hyperboloid fitted to the barrel in TIM. It is typical of the results obtained with the other eight barrels, summarized in Table 1. The fit is generally very good near the equatorial planes of the hyperboloids which always cut through the β-sheets, but deteriorates further away from the equator. It remains acceptable for the purpose of our analysis, however, since the average deviation of a β-strand axis from the orientation of a generating line $\langle\theta\rangle$ is small (5–9°), and comparable to the standard deviations of relevant angular parameters such as

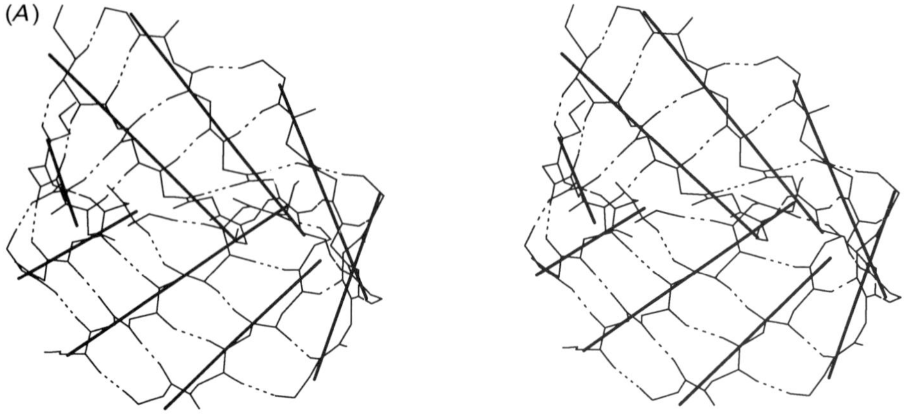

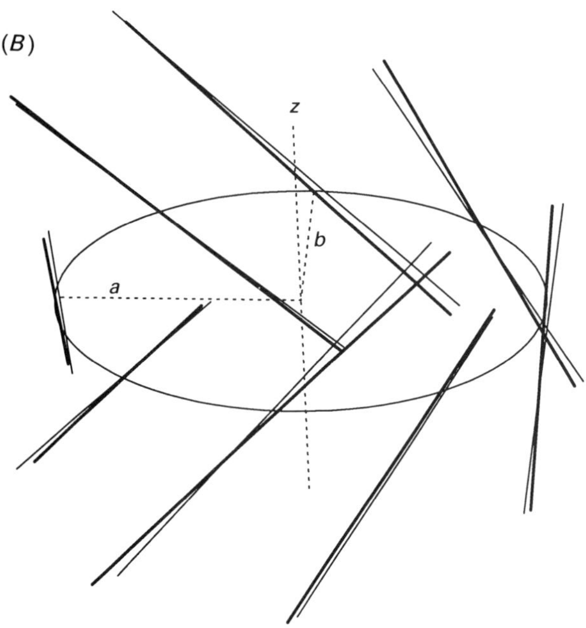

Fig. 1. *Fit of a regular hyperboloid to the β-sheet portion in TIM*

(*A*) Stereo picture of the main-chain of β-strands in TIM (thin lines) and computed least-square axes (bold). H-bonds between strands are drawn in dashes. Limits of β-strands are as specified in [30]. (*B*) Deviations of β-strand axes (thin lines) from the expected orientations of corresponding generating lines (bold) for TIM. The hyperboloid equatorial elliptical cross-section is drawn. Major (*a*) and minor (*b*) semiaxes as well as the hyperboloid axis (*z*), are drawn in dashes. The graphics are generated using the BRUGEL package [58].

T_z (the average angle made by β-strand axes with the barrel axis), or the inter-strand twist (see below). A better approximation to the barrel surfaces could undoubtedly be obtained with a more sophisticated model such as the twisted hyperboloid [29], but such a model would not be as conveniently expressed in terms of strand number or twist, as its simpler variant used here.

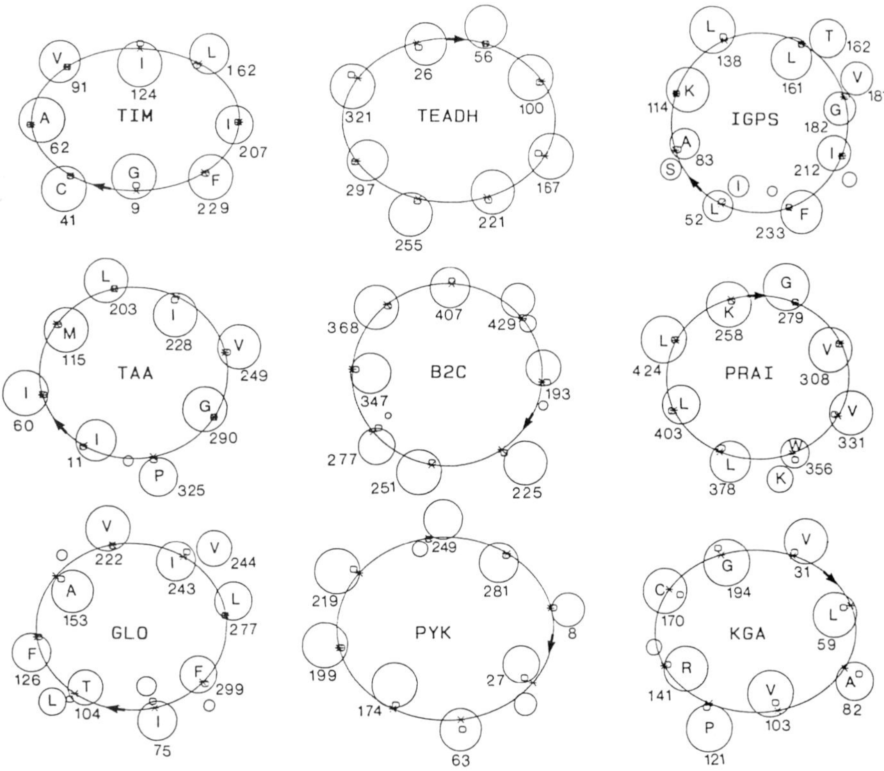

Fig. 2. *Equatorial cross-sections of parallel β-barrels from eight proteins*

Intersection points of least-squares lines passing through β-strands (○) and of generating lines in fitted hyperboloids (×) are shown. *N*-terminii of β-strands are directed towards the viewer and arrows pointing from β-strands *a*–*b* indicate arrangement of strands in the barrels. Large circles represent cross-sections produced by the equatorial plane cutting through the van der Waals spheres of all the barrel C_α atoms. These atoms are marked by their sequence number and, whenever possible, by the one letter amino acid code. Figure from Lasters *et al.* [30] with permission.

Geometrical parameters corresponding to hyperboloids fitted to each of the nine barrels analysed are given in Table 2. $\langle T_z \rangle$ equals 36°, and is virtually invariant among barrels. The *c* parameter is also very similar in different proteins, whereas significant variability occurs in the lengths of the semiaxes *a* and *b*. Consequently, the barrel axial ratios $l = a/b$ vary by as much as 40%, corresponding to a wide range of equatorial shapes: the barrels in cytochrome b_2 and indoglycerol phosphate synthase (IGPS) are nearly circular with $l = 1.04$, and $l = 1.0$, respectively, whereas that of TIM is elliptical with $l = 1.48$. On the other hand, areas of the equatorial cross-sections, computed using the expression $A = \pi a \cdot b$, remain rather constant, with a mean value of 163 Å^2, and a standard deviation of about 10%. For each barrel we also compute the average distance $\langle D \rangle$ between axes of adjacent β-strands and the average angle between these axes (the twist angle) $\langle T_w \rangle$. Both parameters, which are independent of the hyperboloid fit, vary little among barrels. In all 72 pairs of adjacent β-strands in our sample $\langle D \rangle$ has a mean of 4.58 Å and a standard

Table 1. *Geometric parameters of parallel β-barrels*

Protein abbreviations are as follows: triose phosphate isomerase (TIM), taka-amylase (TAA), glycolate oxidase (GLO), triethylamine dehydrogenase (TEADH), flavocytochrome b_2 (B2C), pyruvate kinase (PYK), aldolase (KGA), indoglycerol phosphate synthase (IGPS) and phosphoribosylanthranilate isomerase domain 1 and 2 (PRAI). a, b and c are respectively the major and minor semiaxes and the curvature along the z-axis of the fitted hyperboloid; A is the equatorial surface; $\langle D \rangle$ is the average distance between axes of adjacent β-strands; $\langle T_z \rangle$ is the average angle between β-strand axes and the z-axis of the fitted hyperboloid; $\langle T_w \rangle$ is the average twist angle between the axes of adjacent β-strands, and $\langle \theta \rangle$ is the average deviation between axes of β-strands from the expected orientation of the corresponding generating lines of a hyperboloid of revolution. The standard deviations of D, T_z and T_w appear in parentheses. The hyperboloids are fitted to C_α coordinates of β-strands using the procedure described in [30]. Table from Lasters *et al.* [30] with permission.

	a (Å)	b (Å)	c (Å)	A (Å^2)	$\langle D \rangle$ (Å)	$\langle T_z \rangle$ (degrees)	$\langle T_w \rangle$ (degrees)	$\langle \theta \rangle$ (degrees)
TIM	8.3	5.6	10.2	146	4.4 (0.4)	−35 (5)	−25 (11)	5.2 (2)
TAA	7.5	6.7	10.0	158	4.2 (0.2)	−37 (8)	−28 (6)	8.7 (2)
GLO	7.5	6.6	9.8	155	4.2 (0.6)	−36 (4)	−26 (11)	8.5 (4)
TEADH	8.4	6.4	11.1	170	4.7 (0.5)	−35 (4)	−25 (9)	5.6 (4)
B2C	7.5	7.2	10.1	169	4.6 (0.4)	−37 (5)	−27 (5)	5.2 (3)
PYK	8.6	7.2	11.3	195	5.0 (0.5)	−36 (5)	−27 (7)	6.6 (2)
KGA	8.1	6.4	10.2	163	4.4 (0.9)	−36 (6)	−26 (15)	8.2 (4)
IGPS	7.1	7.1	11.2	158	4.4 (0.4)	−34 (8)	−25 (6)	9.6 (3)
PRAI	7.4	6.4	10.7	149	4.4 (0.3)	−33 (4)	−24 (6)	4.8 (2)

deviation of 0.34 Å, and $\langle T_w \rangle$ has a mean of $-26°$ and a standard deviation of only 8°. By way of comparison, the distribution of twist angles in anti-parallel β-sheets has a mean of $-24°$ and a standard deviation of 17° [34]. The latter is nearly twice as large as the sample analysed here, confirming earlier observations by Salemme and co-workers [35,36] that the surfaces of β-sheets formed by parallel β-strands are more regular than surfaces of β-sheets formed by anti-parallel strands. Moreover, our analysis clearly indicates that earlier conclusions by Chothia & Janin [34], based on the study of anti-parallel β-sheets and stating that barrels cannot be readily formed by the uniform distortion of a regular sheet, do not apply to the type of structures analysed here.

Sheet packing in observed β-barrels

Since the axis of a regular hyperboloid is also a symmetry axis, any portion of the β-sheet surface may be generated from another by a twofold rotation about this axis. The angle between sheets therefore takes the value $\Omega = 2\langle T_z \rangle$. With values of $\langle T_z \rangle$ of $-36°$ as obtained in our barrel sample, Ω is $-72°$ which corresponds to orthogonal sheet packing [33]. To evaluate the contributions of sheet–sheet interactions to the observed barrel structures, a more detailed analysis of sheet packing has been carried out on six β-barrels for which positions for side-chain atoms were available. These are: TIM, TAA, GLO, KGA, IGPS and PRAI (see legend of Table 1 for protein abbreviations). In each case, interactions between different β-sheet pairs, corresponding to

Table 2. *Geometric parameters of circular hyperboloids*

Geometrical parameters of circular hyperboloids computed using a fixed distance $D = 4.5$ Å, and three different twist angles T_w (exp. *a*, *b*, *c*). For each twist angle, listed parameters correspond to all real solutions of eqns (2) and (3). *N*, number of strands in hyperboloid; *R*, hyperboloid radius; Ω, angle between β-strand axes across the hyperboloid, computed as $2 \times T_z$, with $T_z = \arccos [c/(c^2 + R^2)^{0.5}]$, being the angle of a generating line (or β-strand axis) with the hyperboloid axis, and *c* is as explained in the text; *A*, equatorial cross-section; *S*, shear number, is computed using the expression $S = N[(D/a) \tan (T_z)]$, where *a* is the distance between adjacent C_α atoms in a β-strand projected along its axis, and other parameters are as defined above. Table from Lasters *et al.* [30] with permission.

N	*R* (Å)	Ω (degrees)	*A* (Å^2)	*S*
Exp. *a*: $D = 4.5$ Å, $T_w = -26°$				
4	3.3	−37	34	1.9
5	4.0	−45	51	2.9
6	4.9	−53	76	4.2
7	5.9	−63	110	5.9
8	7.1	−72	158	8.0
9	8.5	−82	229	10.9
10	10.3	−93	338	14.7
11	12.9	−106	527	20.2
12	17.1	−121	926	29.2
Exp. *b*: $D = 4.5$ Å, $T_w = -40°$				
4	3.4	−58	37	3.1
5	4.5	−71	62	4.9
6	5.8	−86	106	7.8
7	7.9	−104	198	12.4
8	12.3	−127	479	22.1
Exp. *c*: $D = 4.5$ Å, $T_w = -60°$				
4	3.9	−90	48	5.5
5	6.3	−117	125	11.2

circular permutations of four adjacent β-strands, have been evaluated quantitatively. The procedure consists in computing, for each β-sheet pair, a quantity termed the 'van der Waals contact area', using an analytical algorithm by P. Alard (unpublished work). This quantity is obtained by summation of contact areas of individual atom pairs (computed in the presence of β-sheet residues only) where each member of the pair belongs to a different sheet. For a given pair of atoms the contact area is defined as the portion of the circular cross-section formed by their interpenetrating Van der Waals radii (also called the radical plane) that is not masked by radical planes formed with neighbouring atoms. Care is taken not to include contacts between adjacent β-strands in the computations.

In TIM, consistent with its high axial ratio ($l = 1.48$), one β-sheet pair, composed respectively of strands *defg* and *habc*, interacts more extensively (contact area of 68 Å^2). In the other five barrels whose cross-sections are more circular, contact areas between sheets are significantly smaller (the largest area is 35 Å^2 for the *defg*/*habc* pair in aldolase) with little or no packing preference for any given pair of sheets. For comparison, contact areas between several anti-parallel sheets exhibiting orthogonal packing have been computed by the same procedure using published coordinates [37] and sheet definitions [34]. Intersheet contact areas in alcohol dehydrogenase and tosyl-elastase (respectively 149 and 132 Å^2) are about twice as large as the largest contact area between sheets computed in our sample, along the minor axis in TIM, whereas in *Staphylococcus* nuclease, sheet packing is less extensive (90 Å^2), but still tighter than in TIM. Weaker interactions in orthogonally packed parallel versus anti-parallel sheets may have multiple origins: larger separations between sheets, less strand coiling, difference in amino acid composition, or simply lesser accuracy of the atomic coordinates of the protein sample studied here. Data from a larger sample of refined barrel structures are presently being examined to re-evaluate these results.

The possibility that amino acid composition may play a role in these observations has also been raised. However, comparisons of the amino acid composition of residues involved in sheet packing in the barrel sample analysed in our study (our unpublished work) with that found in anti-parallel β-sheet packing [34], indicate no significant differences. Yet, a closer examination of sheet packing in TIM suggests that an influence from amino acid composition exists. The TIM barrel has a sizeable cavity in its centre [38], lined by small residues, Ala-62 and Gly-9. This cavity will account at least in part (our unpublished work) for the smaller contact area in TIM relative to *Staphylococcus* nuclease where intersheet distance is 11.4 Å, compared with 10.1 Å in TIM.

A detailed analysis of residue packing in the barrel interior has been performed by Lesk *et al.* [39] for three proteins: TIM, glycolate oxidase, and Rubisco. It describes the packing in this region to involve three layers of residues, with each layer containing four residues. The authors differentiate between two classes of packing. In one class, to which Rubisco and TIM belong, the central layer is formed by residues from odd-numbered strands, and in the second class, to which glycolate oxidase belongs, it is formed by residues from even-numbered strands. It is then suggested that each class has evolved independently. The main argument being that evolutionary interconversion between the two classes by a succession of point mutations is extremely unlikely, because it would require reconstruction of the protein cores as well as the active sites. This argument is, however, weakened somewhat by considering that packing between secondary structure elements may be sufficiently non-specific [40] to allow such reconstruction to occur at minimal cost in stability and by recognizing that other interconversion mechanisms, involving for example permutations of entire gene segments, analogous to those recently achieved *in vitro* [27] cannot be ruled out.

Finally, these authors have also performed volume calculations [41] for residues that participate in sheet–sheet interactions in the three structures, and found average residue volumes and packing densities to be similar to those generally found in proteins (C. Chothia, personal communication). These observations, which must of course be verified for a larger set of barrels, are not necessarily inconsistent with the conclusions made above on the basis of sheet contact areas. It is indeed possible that in parallel β-barrels the splaying apart of β-strands is larger than in anti-parallel β-sheets where strands display an appreciable amount of coil. As a result, the contact area between sheets is smaller, although, in the contact region formed by the layers of residues centred at the barrel equator, side-chains may still be optimally packed.

Constraints that govern the topology of regular barrels

To establish the relation between strand number and other key parameters of the β-barrels, and to analyse the constraints that determine the topology of these structures, the hyperboloid equations have been re-parameterized. For simplicity, the case of a circular hyperboloid was considered. The parameters needed to describe it analytically are $a = b = R$ and c, with R being the radius of the equatorial cross-section. These parameters can be readily expressed in terms of the distance D, the twist angle T_w, between two adjacent regularly-spaced generating lines and the number N of generating lines, using the following equations:

$$c^2 = D^2[\cos(T_w) + 1]/[4(1 - \cos(T_w))] \tag{2}$$

$$R^2 = c^2[1 - \cos(T_w)]/[\cos(T_w) - \cos\alpha] \tag{3}$$

where $\alpha = 2\pi/N$ is the angular spacing between two adjacent generating lines. Thus, given T_w, D and N, the hyperboloid is fully determined.

Our analysis has shown all three parameters to be highly conserved in the barrels we studied, suggesting that important constraints may be involved. The interstrand distance D is unquestionably the most strictly constrained parameter—it must remain around 4.5 ± 1 Å to allow for H-bonds to form between strands—whereas the constraints operating on the twist angle T_w and the strand number N are less obvious. To examine this question equations (2) and (3) were solved keeping D fixed at 4.5 Å and using three different values of T_w respectively, $-26°$, $-40°$ and $-60°$. For each of these values N was varied systematically in the range that yields real solutions of the quadratic hyperboloid equations. Parameters of the computed hyperboloids are listed in Table 2.

1. *Relation between strand number and twist angle.* Clearly evident from Table 2 is the strong interdependence of the twist angle T_w and the strand number N. When $T_w = -26°$, the observed value in our barrel sample, a rather wide range of N (4–12) yields solutions to the hyperboloid equation (Table 2 exp. *a*). This range is markedly reduced in hyperboloids with higher twist angles: N ranges from 4 to 8 for hyperboloids with $T_w = -40°$, and only from 4 to 5 for $T_w = -60°$ (Table 2 exp. *b* and exp. *c*). We see, however, that computed hyperboloids span a wider range in N than real β-barrels. This is verified especially for hyperboloids with $T_w = -26°$ and suggests that constraints due to the physical nature of proteins may be at play to further reduce the allowed range in N.

2. *Requirement for optimal packing inside the barrel.* It is reasonable to assume that a stable barrel structure will exclude bulk solvent from the interior and display favourable intramolecular packing interactions. In the eight-stranded barrels, optimal packing is achieved by shielding an entire layer of residues—the central layer—from solvent through interactions with the

layers above and below [39], which also bury side-chains in the process. It is unlikely that barrels with N less than five or greater than eight can achieve this effectively.

For hyperboloids with $N < 5$ and either of the three twist angles of Table 3, cross-sectional diameters are narrower than 8 Å, which is less than the shortest distance between sheets encountered in proteins [34]. With twist angles larger than 60°, cross-sections will increase, but so will the opening of the corresponding barrels. Then the central layer of residues will be less well protected from solvent by the layers above and below, most probably resulting in less stable structures.

On the other hand, hyperboloids with more than eight strands have cross-sections of 16 Å or more, and are therefore wider than the widest hyperboloid in our protein sample. Intersheet distance in the corresponding barrels is therefore too large to foster contacts, and the barrel interior may become accessible to solvent. This should be energetically unfavourable, at least as long as the barrel interior is hydrophobic. The effect of N on the dimensions of the barrel interior is emphasized by its dramatic influence on A, the area of the equatorial cross-section (Table 2, column 4). In hyperboloids with $T_w = -26°$, each additional strand in the barrel corresponds to a 50% increase in A, while in those with $T_w = -60°$ (Table 2*c*), going from four to five strands with nearly triple this area. Moreover, only very few hyperboloids in Table 2 have values of A that fall within 25% of the average of 163 Å^2, computed in our protein sample. It is satisfying to see that, for $T_w = -26°$, only hyperboloids with eight strands have A values in this range. For larger twist angles, $T_w = -40°$ and $-60°$, our model predicts that barrels with respectively seven and five strands should have acceptable values of A. This prediction seems to be born out at least in part. Indeed, the five-stranded barrel in Table 2 exp. *c* ($T_w = -60°$) closely resembles the parallel β-barrel formed around 5-fold axis by the N-terminal part of the VP3 domains in the Rhino virus structure [42]. Using published coordinates [37] and the procedure described above, we determined the following parameters for this barrel: $T_w = -63°$, $R = 5.97$ Å, $D = 3.8$ Å, and $T_z = -62°$. These values are in very good agreement with our predictions. It is worthwhile to add that, as a result of the large twist angle between strands in this structure the barrel is much more open, with c values around 3 rather than 10 as in the eight-stranded barrels computed with $T_w = -26°$. Our calculations suggest that barrels with intermediate twist angles (i.e., $-40°$) should display intermediate characteristics.

3. *Other constraints.* We have seen in the preceding sections that the mathematical constraints imposed by the hyperboloid equations and the requirement for optimal packing inside the barrel are sufficient to explain the properties of observed parallel β-barrels. Other factors such as the angle between sheets across the barrel (Ω), or maintaining a regular arrangement of residues on the barrel surface (a property described by the shear number S [43]) are therefore bound to be of lesser consequence. Indeed, we see (Table 2) that Ω increases with increasing N in circular hyperboloids computed with constant D and T_w, and that it spans a wide range of values ($-40°$ to 127°), all of which have been observed in proteins either in aligned or orthogonal sheet

packing. The role of the shear number S is harder to assess in the framework of our model. S must have even parity for a hyperboloid whose surface is made of a regular β-sheet [43]. We find that only specific values of N correspond to an even shear number in the circular hyperboloids (Table 2). But since the value of S is very sensitive to values of other geometrical parameters such as interstrand distance, only qualitative conclusions can be drawn. We have so far considered only circular hyperboloids. It may be argued therefore that barrels with $N > 8$ could still comply with packing requirements by deviating from a circular shape, in particular since we observe axial ratios of up to 1.48 in our barrel sample. This question has been briefly addressed by extending our analysis to hyperboloids with elliptical cross-sections. We find that the variance in T_w increases drastically with increasing axial ratio: with $N = 10$ and an axial ratio of 3, $T_w = -26.3° \pm 20°$. Highly elliptical barrels will therefore have very twisted corners and be more like double layers or orthogonally packed β-sheets (values of Ω remain between $-70°$ and $-110°$) resembling orthogonal arrangements found in anti-parallel sheets, except that, in the latter case, the highly twisted corners are achieved through a right-angle bend of a long β-strand [34] that provides a covalent link between layers, while in the parallel case such a link does not occur.

Use of the Hyperboloid Model to Design Idealized β-Barrels

Since the hyperboloid model provides a good description of the β-sheet portion of parallel $\alpha\beta$ barrels, its use as a scaffold for the design of idealized β-barrels has been tested [31]. The scaffold-based approach has distinct advantages over other reported methods for computer-aided design of β-sheets which involve the use of virtual atoms [35] or the minimization of objective functions [44]. It allows a systematic investigation of the requirements for main-chain and side-chain conformations leading to a stable β-barrel of predefined geometry, which can moreover be altered in a completely controlled fashion.

The design methodology involves two main steps. The first step consists in defining the geometry of the scaffold, and the second step in mapping β-strands on to the scaffold. To define the scaffold geometry, guidelines are provided from the above described analysis of barrel design features with, in addition, a number of simplifying assumptions. In the first application of this approach, the simplest and most symmetrical barrel geometry, that with a circular cross-section, has been considered [31]. This is justified by the fact that the observed variability in equatorial axial ratios of known barrels is as yet not understood. In this case only two parameters are needed to fully define the barrel. They are R and c, the barrel radius and the curvature of the hyperboloid surface along the barrel axis, respectively. Considering that the equatorial surface in known barrel structures is closely conserved at about 160 Å^2 (see Table 1), the barrel radius is assigned the corresponding value of 7.1 Å. The value of c is then computed from eqn. (3) using this value of R together with $N = 8$ and with $T_w = -26°$, the average value of the twist angle between β-strands observed in known barrel structures (Table 1). By virtue of equations (2) and the expression for T_z (see legend of Table 2), this also determines the values of other relevant parameters such as the interstrand distance D (4.51 Å) and the angle T_z ($-36°$) between β-strand axes and the barrel axis, which are also well conserved in barrels of known structure.

In the second step, ideal β-strands each containing seven residues, so as to allow a central layer of residues to be positioned at the barrel equator, are built using standard bond length and bond angles and assigning identical values to the ϕ, ψ dihedral angles of all residues. Helical parameters, including the β-strand axis, needed to compute the rigid body transformations which will map the strands on to the scaffold, are then readily computed from the atomic coordinates [45].

Analysis β-Sheet Closure Requirements

We then proceed to analyse the requirements with respect to backbone geometry (ϕ, ψ angles) resulting from the formation of a closed β-sheet surface in which a regular pattern of H-bonds is formed. For that purpose, a system-

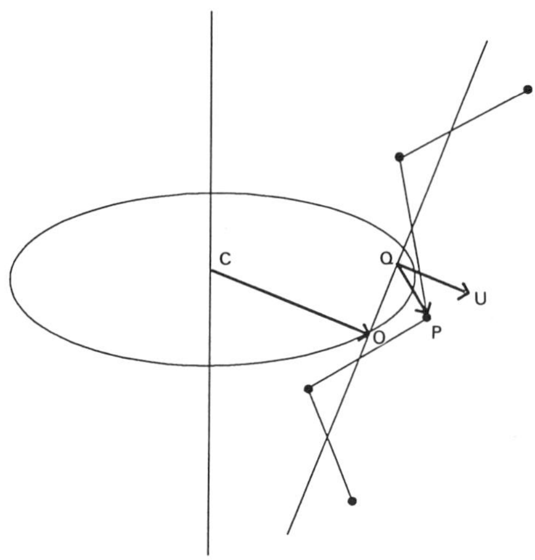

Fig. 3. *Definition of translation and rotation about the β-strand axis situated on the hyperboloid surface*

Q is the projection on the β-strand axis of a reference C_α atom, P. O is the intersection point of the β-strand axis with the barrel equator. C is the centre of the barrel. QU is a vector parallel to CO. Considering the β-strands as rigid bodies, each strand once mapped on the scaffold surface has still two degrees of freedom: translation and rotation about the scaffold axis. To measure the translation (s) and rotation (q) parameters, a reference C_α atom is selected in each β-strand (for convenience the middle C_α in each β-strand is used) and the following procedure is applied: (a) project orthogonally P, the reference C_α coordinates, on the corresponding scaffold axis yielding the point Q; (b) determine the length (l) of OQ, where O is the intersection of the scaffold axis with the barrel equator. The translation parameter for strand i is then defined as: $s_i = 1$ if P is situated above the barrel equator; $s_i = -1$ if P is situated below the barrel equator; (c) compute θ from the clockwise rotation around the scaffold axis from QU to QP, where QU is a vector parallel to the perpendicular between the barrel axis and the scaffold axis (for a hyperboloid of circular cross-section this perpendicular is defined by CO, where C is the centre of the hyperboloid). From Lasters *et al.* [31] with permission from Wiley–Liss Inc.

atic search for optimal H-bonding interactions was performed as a function of backbone dihedral angles and rigid body degrees of freedom of the β-strands: a translation T along the strand axis, and a rotation θ about this axis, as described in Fig. 3.

Owing to the fact that in the idealized barrel β-strands are tilted by $T_z = -36°$ with respect to the barrel axis, residues of the β-sheet surface are organized in layers parallel to the barrel equator, with the central layer almost always lying in or close to the equatorial plane, and residues in the layers pointing alternately inside and outside the barrel surface, as illustrated in Fig. 4. Consequently, the barrel displays a 4-fold symmetry about the barrel axis, with two strands in the asymmetric unit.

Based on these observations the translational parameters of the β-strand have been fixed by placing the central C_α atom of each strand in the plane of the equator. Then, to conserve 4-fold symmetry it is required that rotations

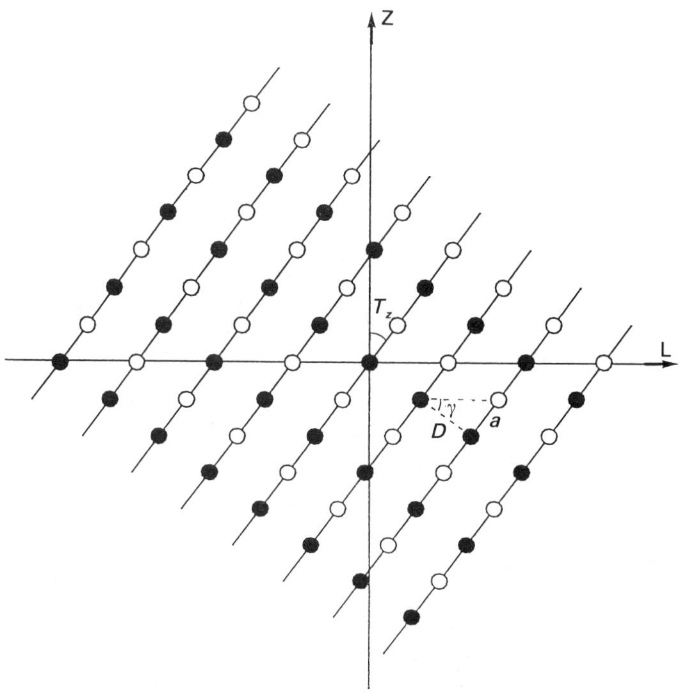

Fig. 4. *Cylindrical projection of the β-sheet surface in an idealized barrel*

The horizontal (L) and vertical (Z) lines are the trace of the equatorial plane and the projection of the barrel axis, respectively. The diagonal lines are projections of β-strand axes, with small circles representing projections of C_α positions of residues pointing inside the barrel (filled circles), or outside the barrel (empty circles). T_z is the angle made by β-strand axes with the barrel axis; a and D, are the distances between adjacent C_α atoms along a β-strand, and between two β-strand axes, respectively. Taking $a = 3.26$ Å, and $D = 4.5$ Å, which is representative of our protein sample and of proteins in general [50], the γ angle equals 36°, precisely the value (but not the sign) of T_z. As a result of this remarkable coincidence, the direction of the diagonal of the β-sheet lattice is parallel to the equatorial plane. Thus the direction of longest distance between neighbouring C_α atoms on the sheet surface coincides with that of highest curvature.

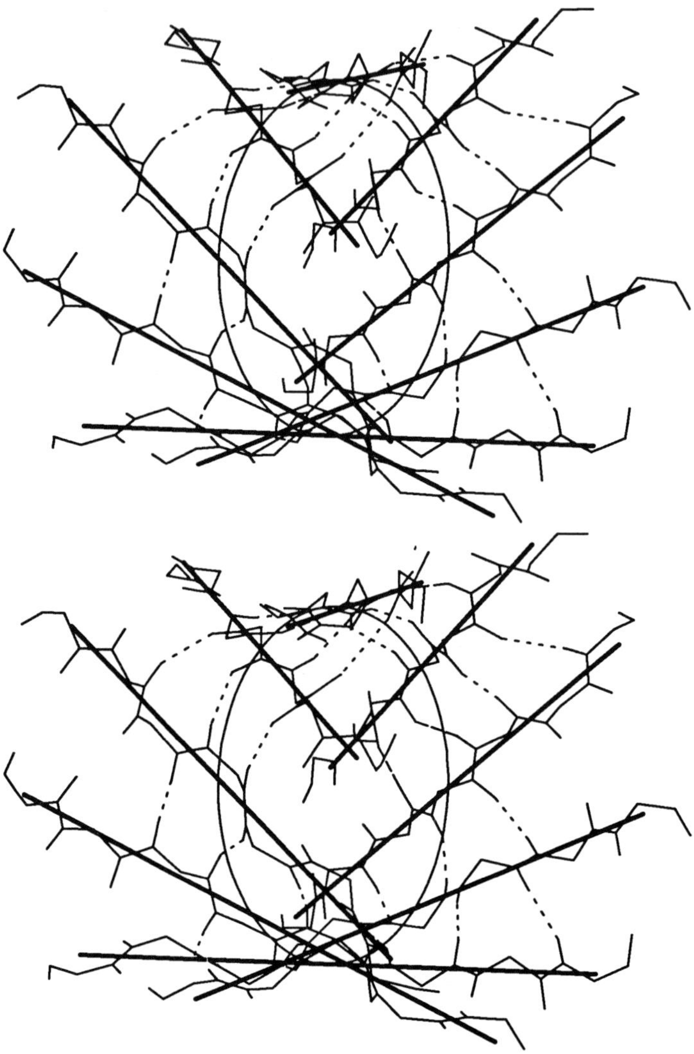

Fig. 6. *Wire diagram representation, in stereo, of the main-chain arrangement and interstrand H-bonds (in dashes) in an idealized β-barrel ($\phi = -114°$, $\psi = 132°$)*

The scaffold used for building the barrel is shown in bold. Graphics are generated using BRUGEL.

about the strand axes, the only remaining rigid-body degree of freedom, be the same in every other strand, leading to independent parameters, Θ_{odd} and Θ_{even}, which are the rotations about axes of odd and even strands, respectively. As a result, a total of only four independent parameters ϕ, ψ, Θ_{odd} and Θ_{even} are considered. Note that strictly speaking, the 4-fold symmetry also allows the use of two different sets of ϕ, ψ pairs, $(\phi, \psi)_{odd}$ and $(\phi, \psi)_{even}$. A unique set of values was used here for all strands to save computer time. The search then consisted in varying ϕ, ψ in the interval of [$-150°$, $-40°$] and [$40°$, $170°$], respectively, in steps of 4°. For each ϕ, ψ pair, a systematic search was done to determine the values of Θ_{odd} and Θ_{even} that minimize a hydrogen-bonding energy function which includes van der Waals and electrostatic contributions [46].

Fig. 5 displays a contour map showing the ϕ, ψ region leading to the most favourable H-bonding energy in the idealized symmetry barrel considered here. The minimum corresponding to the value of -106 kcal mol^{-1} occurs near $\phi = -114°$ and $\psi = 132°$. The resulting β-barrel superimposed on its scaffold is illustrated in Fig. 6 (see facing page). Analysis of the model shows that it yields a regular pattern of H-bonds, as well as a regular stagger between strands, demonstrating that it is possible to achieve such an arrangement in a closed β-sheet while imposing 4-fold symmetry. In a previous construction [35] obtained with dihedral angles $\phi = -102°$, $\psi = 129°$, irregular stagger

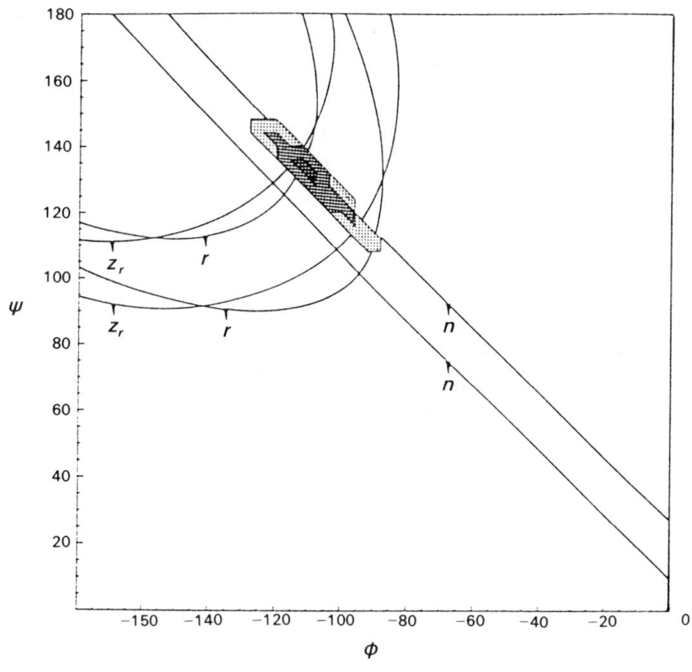

Fig. 5. *Hydrogen bond energy contour plots for idealized β-barrels*

Contouring levels are -104 kcal mol^{-1}, -96 kcal mol^{-1}, and -88 kcal mol^{-1}. The lowest contour level has the darkest shading. Also shown are the contour plots for the helical parameters $n \in [-2.3, -2.1]$, $r \in [0.9$ Å, 1.1 Å$]$, $z_r \in [3.2$ Å, 3.4 Å$]$, which are defined in the text.

was observed, possibly due to H-bonding constraints resulting from the use of a larger barrel diameter (approximately 16 Å), compared with the value of 14.2 Å used here, which has been inferred from a careful analysis of known β-barrel structures. This indicates that the meticulous design of the scaffold may be a crucial factor.

Next, it can be readily seen from Fig. 5, that the low H-bonding energy region coincides with the region in the Ramachandran map corresponding to values of n between -2.1 and -2.4, with n being the number of residues per turn in the β-strand. Hence, the generated β-strands are left-handed helices, in agreement with previous work showing that only left-handed helices are compatible with the observed left-handed twist between β-strand axes [44].

On the other hand, it is observed that this constructed closed β-sheet contains only three good H-bonds between adjacent strands, with average distance of 2.0 ($\pm$0.19) Å between main-chain hydrogen and carbonyl oxygens, with the middle H-bond being sectioned by the equatorial plane. A similar upper limit of less than four H-bonds between adjacent strands in idealized β-barrels has been previously encountered [35], although analysis of real parallel β-barrels (our unpublished work) clearly shows that on average more H-bonds are formed. Indeed even in glycolate oxidase [10], which in our barrel sample displays the smallest number of H-bonds, on average 3.6 H-bonds can be computed between adjacent strands using an earlier partially refined coordinate set. Hence, means to increase the number of H-bonds in idealized β-barrels must be devised. As already pointed out [35], this would require introducing more complex ϕ, ψ patterns to define the β-strand conformation, the effect of which would invariably be to increase strand coil. To verify whether more extensive H-bonding could be achieved in the idealized barrel built here, β-strands were generated as polyalanine chains, and 250 cycles of conjugate gradient minimization were applied to all idealized barrels whose ϕ, ψ dihedral angles were within the two lowest contour levels of the H-bonding energy plot of Fig. 5. Verifying that the symmetry of the β-barrel has been conserved upon minimization, analysis of the minimized structures has been carried out only on three adjacent strands. It showed that in the resulting structures, the parameters of which are summarized in Table 3, the number of H-bonds between adjacent strands increases to an average of 5.4 Concomittantly, some coiling of the main-chain is observed. From detailed comparison of the original and minimized structures it is seen that larger displacements occur near strand extremities, whereas the middle portions are least distorted. In spite of these distortions, the mean ϕ, ψ angles of the minimized structures remain fully confined to the same regions of the Ramachandran map as the backbone angles of barrels built from straight β-strands. These results show that the systematic scaffold-based design procedure described here leads to a proper limitation of backbone dihedral angle space compatible with good H-bonding in eight-stranded regular barrel structures.

It has furthermore been verified that these conclusions, reached for a polyalanine barrel, also hold for a polyvaline β-barrel model built using an identical scaffold. This is an important test of the approach since β-branched bulky hydrophobic residues such as Val and Ile are much more abundant in β-sheets

Table 3. *Energy minimization of idealized β-barrels built using the scaffold*

Conjugate gradient energy minimization of all 23 β-barrels having their φ, ψ dihedral angles situated within the two lowest H-bonding energy contours of Fig. 5. Columns 2, 3 and 4 characterize the idealized β-barrels built from straight β-strands: columns 2 and 3 list respectively the φ and ψ dihedral angles and column 4 gives the number of hydrogen bonds between three adjacent β-strands. All β-barrels were subjected to 250 cycles of conjugate gradient minimization. Columns 5–10 present an analysis of the energy minimized structures. This analysis is carried out on three adjacent β-strands. Columns 5 and 6 show respectively the average φ and ψ dihedral angles with the standard deviations in parentheses; column 7 specifies the number of H-bonds; column 8 gives the non-bonded energy (Nbo); columns 9 and 10 list the r.m.s. displacements (expressed in Å) between the energy minimized and the original structures, for all main-chain atoms (RMS all) and main chain atoms of residues corresponding to the three central layers in the β-barrel [39] (RMS cl). From Lasters *et al.* [31] with permission from Wiley–Liss Inc.

No.	φ	ψ	H-bonds	⟨φ⟩	⟨ψ⟩	H-bonds	Nbo (kcal mol^{-1})	RMS all	RMS cl
	Barrels with straight β-strands			Energy minimized barrels					
1	−98	116	8	−110 (28)	125 (25)	12	−56	0.84	0.60
2	−106	128	7	−106 (22)	124 (25)	10	−62	0.68	0.47
3	−102	120	8	−107 (25)	123 (26)	12	−60	0.84	0.62
4	−98	120	8	−106 (30)	124 (24)	10	−57	0.78	0.59
5	−106	120	7	−105 (21)	122 (25)	12	−57	0.87	0.62
6	−110	124	7	−109 (22)	126 (20)	12	−61	0.83	0.60
7	−106	124	8	−105 (20)	123 (25)	12	−62	0.83	0.54
8	−102	124	7	−105 (28)	122 (26)	10	−57	0.77	0.60
9	−110	128	7	−107 (23)	124 (30)	10	−61	0.90	0.62
10	−114	128	7	−111 (19)	128 (19)	12	−63	0.81	0.57
11	−118	132	7	−115 (17)	132 (17)	11	−59	0.70	0.50
12	−114	132	7	−108 (20)	126 (23)	11	−62	0.80	0.56
13	−110	132	7	−107 (22)	126 (23)	10	−63	0.68	0.54
14	−106	132	8	−102 (20)	130 (19)	9	−61	0.69	0.55
15	−122	136	6	−113 (19)	131 (25)	11	−55	0.73	0.45
16	−118	136	6	−108 (24)	127 (29)	12	−63	1.00	0.63
17	−114	136	7	−111 (25)	128 (27)	10	−59	0.91	0.67
18	−110	136	6	−109 (22)	133 (17)	9	−56	0.68	0.57
19	−122	140	6	−112 (22)	132 (28)	12	−62	0.88	0.51
20	−118	140	6	−111 (23)	131 (25)	11	−61	0.88	0.65
21	−114	140	6	−109 (24)	133 (19)	9	−55	0.70	0.59
22	−126	144	6	−113 (25)	131 (29)	12	−62	0.94	0.55
23	−122	144	6	−115 (18)	137 (21)	10	−63	0.76	0.55

than the small Ala side-chain which is preferentially found in α-helices. The building procedure and the systematic search for optimal H-bonding was thus repeated in the presence of valine side-chains, built using dihedral angles $\chi^1 = 177°$, $\chi^{12} = 55°$, $\chi^{22} = 65°$, as described by Chou *et al.* [48] in a study of twisted β-sheets. The φ, ψ values corresponding to the minimum energy structure were found to be −96° and 116°, respectively. These values fall within the distribution of φ, ψ pairs shown in Table 1, indicating that the scaffold can be used to design β-barrels from other than only polyalanine strands. It will require further systematic analyses to determine actual sequence templates that can be harboured in the barrel as built here.

While we see that the scaffold-based construction method is able to yield regular β-barrels with an H-bonding pattern and backbone angles close to those found in real β-barrel structures, an important question to ask is whether forming a closed regular β-sheet surface imposes specific constraints on the conformation of the polypeptide backbone which are different from those in open β-sheets. This question is of particular interest, since it might have important implications on the folding process of β-barrel structures. It is, however, difficult to answer it by analysing ϕ, ψ angles of real β-barrels, since their values show considerable scatter [49]. An alternative approach is to compare values of backbone angles in the idealized β-barrels built here, with average ϕ, ψ, values derived from an exhaustive repertoire of β-strands found in known protein structures. These values characterize an average β-strand that can be described by its helical parameters n (number of residues per turn), r (radius) and z_r (rise per residue along the β-strand axis). Since any pair of ϕ, ψ dihedral angles specifies an ideal helix, the helical parameters of which can be readily computed [32,50], a contour plot can be made in ϕ, ψ space delimiting regions compatible with average β-strand helical parameters [50, 56]. Fig. 5 shows such plots for $n \in [-2.3, -2.1]$, $r \in [0.9$ Å, 1.1 Å$]$, $z_r \in [3.2$ Å, 3.4 Å$]$, superimposed on the H-bonding energy contours. Clearly, the ϕ, ψ requirements to optimize H-bonding in our idealized β-barrel are compatible with average β-strand geometry as can be seen from the overlap of the low H-bond energy region with that limited by the intersection of the n, r and z_r contour plots.

On the Relations between Sequence and Structure Motifs in αβ Barrels

To understand fully the factors that determine the $\alpha\beta$ barrel fold, the role of the amino acid sequence in determining the observed structure must be determined. This is a notoriously difficult task, and has not as yet been achieved for any known protein fold. Based on the knowledge of the amino acid sequence and the structures of evolutionarily related proteins it has however been possible to derive consensus sequence patterns, also termed templates or profiles, that characterize to various extents the corresponding protein family [51–54]. Such templates can then be used as probes to detect functional or evolutionary relatedness—and consequently also structural resemblance—between proteins that display very low sequence identity. Since the proteins that adopt the $\alpha\beta$ barrel fold display great sequence variability, and possibly also due to high internal symmetry of the structural motif, deriving sequence templates which characterize this fold is a very difficult task.

Recently, Claessens *et al.* [55] rebuilt the backbone of TIM using fragments of variable lengths extracted from a database of known protein structures, while being guided by the crystallographically determined C_α positions. Their study seemed to suggest (see Fig. 7) that $\beta\alpha$ loops, connecting the *C*-terminii of β-strands to the *N*-terminii of α-helices, contain less common folding motifs than the rest of the chain, since they could only be fitted with short contiguous

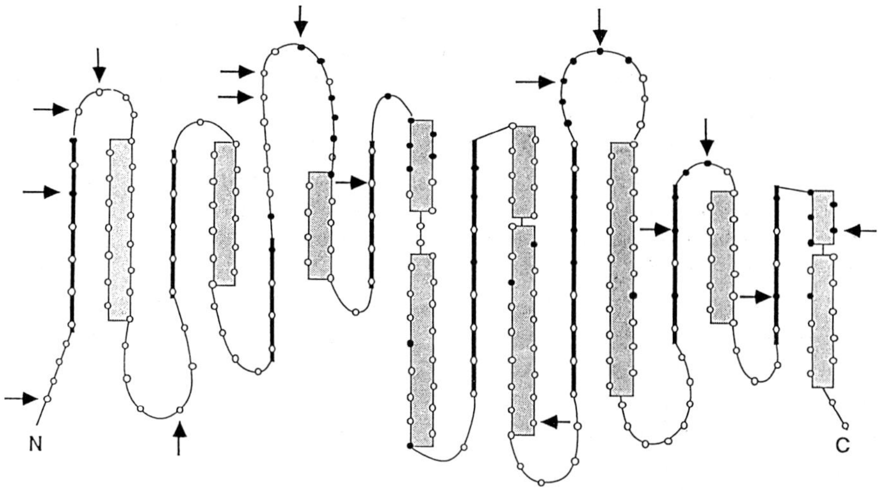

Fig. 7. *Distribution of 'short fragments' in the constructed backbone of TIM*
α-Helices and β-strands are represented by shaded rectangles and filled bars, respectively. Arrows indicate the position of the first residue in the so-called 'short fragments' obtained from the backbone construction procedure previously described [55], and defined as those containing at most six residues. From Taylor [52] with permission from Wiley–Liss Inc.

polypeptide fragments from the database. On the other hand, $\alpha\beta$ loops connecting *N*-terminii of β-strands to *C*-terminii of helices could be fitted by longer continuous segments from the database. This seems to agree well with the observation that in $\alpha\beta$ proteins, and in particular in $\alpha\beta$ barrels, $\beta\alpha$ loops are generally much longer and more variable in length than $\alpha\beta$ loops and sometimes contain entire structural domains. Another remarkable observation is that active sites in $\alpha\beta$ barrel proteins are always located in the *C*-terminal portion of the β-strands and in $\beta\alpha$ loops [24]. All this suggests that $\beta\alpha$ loops may be shaped mainly to comply with more functional requirements, while $\alpha\beta$ loops may be a good place to look for conserved structure and sequence patterns that may characterize basic structural and folding constraints. An exhaustive analysis of $\alpha\beta$ loops in the six barrel structures for which we had atomic coordinates has been carried out (Lasters, I., Scherlink, J. P., Pio, F. & Wodak, S. J., unpublished work). The analysis consisted in comparing the structures of all the $\alpha\beta$ loops in these structures with each other, and to all other $\alpha\beta$ loops in a database containing 75 known protein structures. This was performed using the 'spare parts' backbone building approach previously developed [55]. Preliminary results show that a subset of the $\alpha\beta$ loops in the six barrels analysed adopt characteristic conformations and sequence patterns previously identified in other $\alpha\beta$ proteins as $\alpha\beta_1$ and $\alpha\beta_3$ loops by Edwards *et al.* [32]. We see (Table 4) that all the barrels analysed, except for pyruvate kinase, contain at least one $\alpha\beta_3$ motif, and two proteins, xylose isomerase and glycolate oxidase, contain several such motifs. The $\alpha\beta_1$ motifs are less common and occur only in four of the examined proteins, once in each structure (Table 5). It has moreover been verified (our unpublished work) that optimal coordinate superposition [43] for any two pairs of $\alpha\beta$ barrels also brings about

Table 4. *$\alpha\beta_3$ loops in parallel $\alpha\beta$-barrel proteins*

Summary of the $\alpha\beta_3$ loops located in the α/β barrel proteins of this study. The first column lists the protein, using abbreviations given in legend of Table 1. The second column gives the location of the loops in the barrel, where the order of the β-strands and the α-helices along the polypeptide chain is designated as a–h, respectively. The third column gives the sequence number of the first and last residues defining the structure/sequence motif. The fourth column, designated 'sequence pattern', gives the consensus pattern on top, with H and H* designating hydrophobic residues, and X, any residue. The sequences corresponding to the $\alpha\beta_3$ loops are listed below. The fifth column, (ϕ), indicates the sign of the ϕ angle of the residue in the fifth position in the motif, which is most often occupied by a glycine. The sixth column gives the r.m.s. deviations of backbone atoms of the $\alpha\beta_3$ loop relative to the $\alpha\beta_3$ loop in chicken TIM, after coordinate superposition. XI, xylose isomerase.

β-Barrels proteins	Loop ($\alpha_X\beta_X$)	Sequence	Sequence pattern									ϕ	r.m.s. (Å)
			H	X	X	X	**G**	**A**	X	X	**H***		
TIM													
Chicken	$\alpha_c\beta_d$	83–91	**I**	K	D	I	**G**	**A**	A	W	**V**	+	0.00
Yeast	$\alpha_c\beta_d$	83–91	**I**	K	D	V	**G**	**A**	K	W	**V**	+	0.31
Trypanosomes	$\alpha_c\beta_d$	83–91	**L**	K	D	F	**G**	V	N	W	**I**	+	0.47
XI	$\alpha_c\beta_d$	126–134	**G**	A	E	L	**G**	**A**	K	T	**L**	+	0.57
	$\alpha_a\beta_b$	43–51	**L**	A	E	I	**G**	**A**	Y	G	**I**	+	0.79
TAA	$\alpha_a\beta_b$	52–60	**I**	Q	G	M	**G**	F	T	A	**I**	+	0.80
IGPS	$\alpha_c\beta_d$	130–138	Q	A	D	A	C	L	L	M	**L**	−	0.84
GLO	$\alpha_c\beta_d$	144–152	**A**	E	R	A	**G**	F	K	A	**I**	+	0.60
	$\alpha_e\beta_f$	234–242	**A**	V	Q	H	**G**	**A**	A	G	**I**	+	1.07
	$\alpha_g\beta_h$	290–298	**A**	L	A	L	**G**	**A**	P	G	**V**	+	0.87
PRAI	$\alpha_a\beta_b$	271–279	**A**	Y	D	A	**G**	**A**	I	Y	**G**	+	1.02

Table 5. *$\alpha\beta_1$ loops in parallel $\alpha\beta$-barrel proteins*

Summary of the $\alpha\beta_1$ loops located in the α/β barrel proteins of this study. For all other details, see legend of Table 4.

β-Barrels proteins	Loop ($\alpha_X\beta_X$)	Sequence	Pattern								ϕ	r.m.s. (Å)
			H	X	X	X	**G**	X	X	**H***		
TIM												
Chicken	$\alpha_d\beta_e$	116–123	**A**	L	A	E	**G**	L	G	**V**	+	0.0
Yeast	$\alpha_d\beta_e$	116–123	**A**	L	G	Q	**G**	V	G	**V**	+	0.31
Trypanosomes	$\alpha_d\beta_e$	116–123	**A**	V	A	S	**G**	F	M	**V**	+	0.55
IGPS	$\alpha_d\beta_e$	153–160	**A**	H	S	L	E	M	G	**V**	+	0.58
GLO	$\alpha_a\beta_b$	96–103	**A**	S	A	A	**G**	T	I	M	+	0.80
XI	$\alpha_d\beta_e$	159–176	S	E	D	R	**G**	Y	G	**L**	+	0.99

spatial superposition of equivalent $\alpha\beta_3$ and/or $\alpha\beta_1$ loops. This suggests that the location of these loops in the context of the $\alpha\beta$ barrel is conserved.

Discussion

Parameters that govern parallel β-barrel structures in proteins have been analysed in the context of a model of a regular hyperboloid. The features of these structures are shown to arise primarily from two factors: the right-handed twist of the β-sheet, which governs both the topology of the β-sheet surface and its stability through interstrand H-bonds, and from the requirement to exclude solvent from the barrel interior, which puts severe constraints on barrel dimensions. Barrels which make up the protein core, as in the nine enzyme structures analysed here, appear to tolerate little variability in these two major factors resulting in a remarkable conservation of key geometrical parameters, such as interstrand twist and the strand number ($N = 8$). Our analysis suggests that these barrels represent optimal solutions to the problem of forming regular hyperboloids from a β-sheet. In agreement with this hypothesis is the following additional observation. Adjacent C_α atoms in neighbouring strands lying near or at the equatorial plane, where the surface curvature is highest, point alternately inside and outside the barrel as illustrated in Figs. 2 and 4. This corresponds to a situation of optimal side-chain packing and is a direct consequence of the value of one key parameter: the orientation of strand axes relative to the barrel T_z, which in turn depends on the strand twist T_w (see legend of Fig. 4 for further details). Barrels with T_z significantly different from the values computed in our barrel sample and in the seven- or eight-stranded circular barrels of Table 2 (exp. *a*), will deviate from this ideal packing arrangement and possibly result in less stable structures.

Our study also predicts the existence of other optimal barrel structures with higher twist angles and fewer strands. One of these, a five-stranded parallel β-barrel, actually occurs in the rhinovirus structure. It is interesting to note that this barrel is formed by the N-terminal segment of five polypeptide chains belonging to the VP3 subunits, which come together at the quasi 5-fold axis. The segments have identical and rather symmetric, amino acid sequence with prolines at the ends of each β-strand, and there seems to be less extensive H-bond interactions between strands in the barrel than in the structures analysed here.

While further analysis is clearly needed to confirm our suggestions, our study provides for the first time, a formal description of the relationships between β-sheet twist, strand number and barrel dimensions, that could serve as useful guidelines in studies of protein folding and design. It is shown that the hyperboloid model can serve as a scaffold on to which ideal β-barrels can be designed. Furthermore, this methodology, which can be qualified as a minimalist approach [57], was also applied to study constraints imposed on backbone conformations within the closed β-sheet of an eight-fold $\alpha\beta$ barrel. That study suggested that barrel closure did not impose marked restrictions on main-chain conformation other than those observed for average β-strands in proteins, indicating that folding of these structures does not require global main-chain readjustment upon barrel closure.

Finally, a preliminary report on the analysis of $\alpha\beta$ loops in six barrel structures is presented. It shows that a subset of these loops in each protein adopts conformations and amino acid patterns characteristic of $\alpha\beta_1$ and $\alpha\beta_2$ loops previously identified in other proteins. This might possibly indicate that TIM barrels or their ancestor(s) have evolved by assembly of smaller modules. In line with this hypothesis is the observation that the two $\alpha\beta_3$ loops in xylose isomerase from *Actinoplanes missouriensis* (sequence unpublished) display notable sequence homology (Table 4). Clearly however, more of the known $\alpha\beta$ barrel structures and sequences need to be analysed to further probe into this question.

We thank Drs C. Branden, H. Jansonius, S. Matthews, D. C. Phillips and A. Tulinsky for the gift of atomic coordinates and acknowledge the help of M. Bardiaux, Ph. Delhaise and Dr M. De Maeyer in using the BRUGEL package. The help of Ph. Berthet with the computer systems and of S. De Moor with preparing the manuscript is also gratefully acknowledged. It was Dr M. Rossmann who attracted our attention to the barrel in the rhinovirus structure.

References

1. Richardson, J. S. (1981) *Adv. Protein Chem.* **34**, 167–339
2. Janin, J. & Wodak, S. (1985) *Prog. Biophys. Mol. Biol.* **42**, 21–78
3. Levitt, M. & Chothia, C. (1976) *Nature (London)* **261**, 552–557
4. Banner, D. W., Bloomer, H. C., Petsko, G. A., Phillips, D. C., Pogson, C. I. & Wilson, A. I. (1975) *Nature (London)* **255**, 609–614
5. Muirhead, H., Clayden, D. A., Parford, D., Lorimer, C. G., Fothergill-Gilmore, L. A., Schilz, E. & Schmitt, W. (1986) *EMBO J.* **5**, 475–481
6. Lim, L. W., Shamala, N., Mathews, F. S., Steenkamp, D. J., Hamlin, R. & Xuong, N. (1986) *J. Biol. Chem.* **261**, 15140–15146
7. Mavridis, I. M., Hatada, M. H., Tulinsky, A. & Lebioda, D. L. (1982) *J. Mol. Biol.* **162**, 445–458
8. Matsuura, Y., Kusunoki, M., Harada, W., Tanaka, N., Iga, Y., Yasuoka, N., Toda, H., Narita, K. & Kakudo, M. (1980) *J. Biochem.* **87**, 1555–1558
9. Buisson, G., Duee, E., Haser, R. & Payan, F. (1987) *EMBO J.* **6**, 3909–3916
10. Linqvist, Y. & Branden, C. I. (1985) *Proc. Natl. Acad. Sci. U.S.A.* **82**, 6855–6859
11. Schneider, G., Lindqvist, Y., Andersson, I., Knight, S., Branden, C. I. & Lorimer, G. (1986) *EMBO J.* **5**, 3409–3415
12. Xia, Z. X., Shamala, N., Bethge, P. H., Lim, L. W., Bellamy, H. D., Xuong, N. H., Lederer, F. & Mathews, F. S. (1987) *Proc. Natl. Acad. Sci. U.S.A.* **84**, 2629–2633
13. Carrel, H. L., Rubin, B. H., Hurley, T. J. & Glusker, J. P. (1984) *J. Biol. Chem.* **259**, 3230–3236
14. Hendrick, K., Collyer, C. A. & Blow, D. M. (1989) *J. Mol. Biol.* **208**, 129–157
15. Rey, F., Jenckins, J., Janin, J., Lasters, I., Alard, Ph., Claessens, M., Matthyssens, G. & Wodak, S. J. (1988) *Proteins* **4**, 165–172
16. Hoffmann, B. F., Bender, H. & Schulz, G. (1990) *J. Mol. Biol.* in the press
17. Hyde, C. C., Ahmed, S. A., Padlam, E. A., Miles, E. W. & Davies, D. D. (1988) *J. Biol. Chem.* **263**, 17857–17871
18. Reference deleted
19. Priestle, J. P., Grutter, M. G., White, J. L., Vincent, M. G., Kania, M., Wilson, E., Jardetzky, T. S., Kirschner, K. & Jansonius, J. N. (1987) *Proc. Natl. Acad. Sci. U.S.A.* **84**, 5690–5694
20. Goldman, A., Ollis, D. L. & Steitz, T. A. (1987) *J. Mol. Biol.* **194**, 143–153
21. Rouvinen, J., Bergfors, T., Teeri, T., Knowles, J. K. C. & Jones, T. A. (1990) *Science* **249** 380–386
22. Lebioda, L. & Stec, B. (1988) *Nature (London)* **333**, 683–686
23. Lebioda, L., Stec, B. & Brewer, J. M. (1989) *J. Biol. Chem.* **264**, 3685–3693
24. Fraber, G. K. & Petsko, G. A. (1990) *Trends Biochem. Sci.* **15**, 228–234
25. Longberg, N. & Gilbert, W. (1985) *Cell (Cambridge, Mass.)* **40**, 81–90
26. Luger, K., Hommel, U., Herold, M., Hofsteenge, J. & Kirchner, K. (1989) *Science* **243**, 206–210
27. Luger, K., Szadkowski, H. & Kirchner, K. (1990) *Protein Eng.* **3**, 249–258
28. Goraj, K., Renard, A. & Marcial, J. (1990) *Protein Eng.* **3,** 259–266
29. Novotny, J., Bruccoleri, R. & Newell, J. J. (1984) *J. Mol. Biol.* **177**, 567–573
30. Lasters, I., Wodak, S. J., Alard, Ph. & Van Cutsem, E. (1988) *Proc. Natl. Acad. Sci. U.S.A.* **85**, 3338–3342
31. Lasters, I., Pio, F. & Wodak, S. J. (1990) *Proteins* **7**, 249–256
32. Edwards, M. S., Sternberg, M. J. E. & Thornton, J. M. (1988) *Protein Eng.* **1**, 173–181

33. Chothia, C. & Janin, J. (1981) *Proc. Natl. Acad. Sci. U.S.A.* **78**, 4146–4150
34. Chothia, C. & Janin, J. (1981) *Biochemistry* **21**, 3955–3965
35. Salemme, F. R. & Weatherford, D. W. (1981) *J. Mol. Biol.* **146**, 101–117
36. Salemme, F. R. (1983) *Prog. Biophys. Mol. Biol.* **42**, 95–133
37. Berenstein, F. C., Koetzle, T. F., Williams, G. J. B., Meyer, E. F. Jr, Brice, M. D., Rodgers, J. R., Kennard, O., Shimanouchi, T. & Tasumi, M. (1977) *J. Mol. Biol.* **112**, 535–542
38. Phillips, D. C., Sternberg, M. J. E., Thornton, J. M. & Wilson, I. A. (1978) *J. Mol. Biol.* **119**, 329–351
39. Lesk, A. M., Branden, C. I. & Chothia, C. (1989) *Proteins* **5**, 139–148
40. Chothia, C. (1984) *Annu. Rev. Biochem.* **53**, 537–572
41. Richards, F. M. (1974) *J. Mol. Biol.* **82**, 1–14
42. Arnold, E., Vriend, G., Luo, M., Griffith, J. P., Kamer, G., Erickson, J. W., Johnson, J. E. & Rossmann, M. G. (1987) *Acta Crystallogr. Sect. A. Found. Crystallogr.* **43**, 346–361
43. McLachlan, A. D. (1979) *J. Mol. Biol.* **128**, 49–79
44. Chou, K. C., Pottle, M., Nemethy, G., Ueda, Y. & Scheraga, H. A. (1982) *J. Mol. Biol.* **162**, 89–112
45. Sugeta, H. & Miyazawa, T. (1967) *Biopolymers* **5**, 673–679
46. Reihes, W. E. (1985) Ph.D. thesis, Harvard University, Cambridge, Massachusetts, U.S.A.
47. Weatherford, D. W. & Salemme, F. R. (1979) *Proc. Natl. Acad. Sci. U.S.A.* **76**, 19–23
48. Chou, K. C., Nemethy, G. & Scheraga, H. A. (1983) *J. Mol. Biol.* **168**, 389–407
49. Richardson, J. (1981) *Adv. Protein Chem.* **34**, 167–339
50. Schultz, G. E. & Schirmer, R. H. (1979) *Principles of Protein Structure*, Springer Verlag, New York, Heidelberg
51. Bashford, D., Chothia, C. & Lesk, A. M. (1987) *J. Mol. Biol.* **198**, 327–337
52. Taylor, W. R. (1986) *J. Mol. Biol.* **188**, 233–258
53. Taylor, W. R. (1988) *Protein Eng.* **2**, 77–86
54. Gribskov, M., McLachlan, A. D. & Eisenberg, D. (1987) *Proc. Natl. Acad. Sci. U.S.A.* **84**, 4355–4358
55. Claessens, M., Van Cutsem, E., Lasters, I. & Wodak, S. J. (1989) *Protein Eng.* **2**, 335–345
56. Creighton, T. E. (1984) in *Proteins, Structures and Molecular Properties*, W. H. Freeman, New York
57. DeGrado, W. F., Wasserman, Z. R. & Lear, J. D. (1989) *Science* **243**, 622–628
58. Delhaise, P., Bardiaux, M. & Wodak, S. J. (1984) *J. Mol. Graphics* **2**, 103–106

Biochem. Soc. Symp. **57**, 123–133
Printed in Great Britain

Preparation of Artificial Bifunctional Enzymes by Gene Fusion

LEIF BÜLOW

Department of Pure and Applied Biochemistry, Chemical Centre, P.O. Box 124, S-221 00 Lund, Sweden

Synopsis

Within the cellular framework operate a number of sequentially acting enzyme systems of varying complexity and spatial organization. In some cases they are found as bi- or even multi-functional enzymes, where each single polypeptide chain carries two or more catalytic functions. To mimic these enzymes, a number of different bifunctional enzymes have been prepared by in-frame gene fusion. Thus, β-galactosidase and galactokinase as well as β-galactosidase and galactose dehydrogenase were genetically fused. Different physical and chemical properties of the hybrid proteins have been investigated, such as protein stability, subunit interactions, linker regions and substrate channeling *in vitro* and *in vivo*. Important fields of application for artificial bifunctional enzymes can be found in biochemical analysis, protein purification and metabolic engineering.

Introduction

Enzyme sequences

Enzymes are frequently treated and analysed as isolated and separated biocatalysts. The situation *in vivo* is much more complex. Many enzymes compete with each other for substrates and various effector molecules. For instance, in a metabolic pathway a series of enzymes catalyse consecutive reactions so that the product of the first enzyme is the substrate for the second:

$$A \xrightarrow{\text{Enzyme 1}} B \xrightarrow{\text{Enzyme 2}} C \rightarrow \cdots$$

Furthermore, related enzymes tend to be associated with each other. Such spatial organization can have very different complexity ranging from the very weak interactions found, for instance, between glycolytic enzymes to the formation of multienzyme complexes and multifunctional enzymes carrying two or more active centres on the same polypeptide chain [1]. In general, multienzyme complexes and multifunctional enzymes are more common and more complex in eukaryotes than in prokaryotes. The evolutionary motive for this is likely to be found in the much greater volume of the former organisms. Diffusion problems are thus more likely to occur in eukaryotic cells, although several multifunctional proteins have been found in bacteria. The presence of such highly organized multienzyme systems affects the kinetics of the reaction sequence. The transit time, i.e. the time necessary for the product of the first enzyme to diffuse to the second enzyme, will thus be less in a physically associated enzyme than for two separate enzymes. Also, the time required to reach a new steady state if the initial substrate concentration is changed (the transient time) is influenced. The transient time is accordingly normally found to be less in a bifunctional enzyme. For instance, when analysing the *arom* conjugate, a multifunctional protein carrying five enzymic activities on the same polypeptide chain, the experimentally observed transient time was found to be approximately 1/10 to 1/15 of the theoretically calculated transient time [2]. The intermediate consequently does not appreciably diffuse out into the medium, but moves directly from one catalytic centre to the other. The most important property of multifunctional enzymes is probably the potential of substrate channelling. Once the intermediate is formed it is not generally available to be acted upon by other enzymes outside the complex.

Also, viewed from a more applied viewpoint, consecutive enzymes are often used in biotechnology and biochemical analyses. Beside an obvious need of two or more enzymes to operate in a certain sequence, they can also be used to generate an easily detectable product in an analytical system or to shift an unfavourable equilibrium. Illustrative examples are starch hydrolysis employing the enzymes α-amylase and glucoamylase or the frequently utilized coupling of a dehydrogenase to an enzymic reaction as a final step to enable a simple quantification via NAD(P)H. A special case of sequential reactions is the cyclic reaction where a substance is continuously recycled between two or more enzymes. An interesting example is enzymic coenzyme regeneration where a cofactor is continuously shuffled between an oxidative and a reductive reaction. The industrial importance of this latter type of reaction has increased recently, primarily because of a number of commercially interesting enzymic reactions require expensive coenzymes, like ATP or NAD(P) [3]. Also in the area of bioanalysis, substrate recycling has become extremely valuable, since the sensitivity of the method can be improved considerably. Lactate analysis using co-immobilized lactate oxidase and lactate dehydrogenase can, for instance, improve the sensitivity 1000-fold [4].

Creating proximity

The simplest way a consecutive enzymic reaction can be used is of course to add the native enzymes directly to the reaction mixture. The enzymes are added either simultaneously or one after the other if they would require entirely different conditions, e.g. concerning pH or temperature. However, a number of disadvantages are connected with this method. The cost of the enzymes can be considerable. Furthermore, the stability of the biocatalysts in an enzyme reactor can be limiting. Despite these facts, the method is the most frequently used. A number of alternatives have been described, however, all of which mimic the situation *in vivo* with the formation of a clustering of the enzymes. Thus, an alternative way of utilizing enzymes can be to immobilize them, for instance by coupling them covalently to a carrier material or through confinement in a hollow-fibre reactor. In particular, the first of these alternatives, co-immobilization of two or more enzymes to a matrix, can give the system very attractive properties. In addition to stabilization and reusability of the biocatalyst, very favourable kinetic characteristics can be obtained. For instance, the transient time can often be reduced markedly (cf. Fig. 1) [5,6]. In certain cases, e.g. co-immobilization of malate dehydrogenase and citrate synthetase, it has been demonstrated that even the steady-state rate of the reaction can be increased [7]. The improved kinetic behaviour of the immobilized systems compared with an identical system with soluble enzymes, has been explained by the formation of a favourable 'microenvironment' where the local concentration of the intermediate is higher.

There are also other methods of creating a proximity between two enzymes. A frequently used technique involves chemical cross-linking of the proteins. This normally generates an enzyme aggregation with random orientation of the active centres, but recently more well-defined complexes with a juxtaposition between the active sites have been accomplished [8,9]. A third alternative of generating a proximity between two enzymes is to make an in-frame gene fusion between the corresponding structural genes. The chimeric gene hence encodes a polypeptide chain carrying two or more active centres. This method bears strong resemblance to the proposed model for evolution of multi-

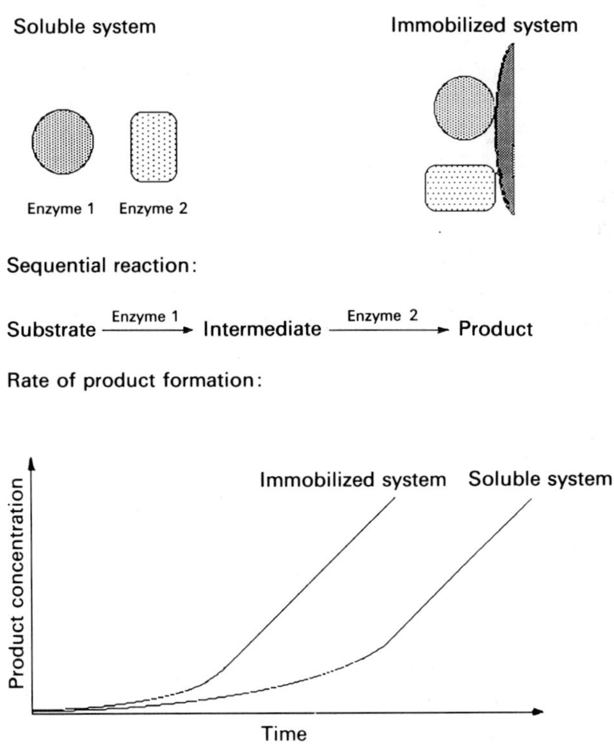

Fig. 1. *Schematic comparison between soluble and immobilized systems*

functional enzymes. Through comparison of DNA sequences from different species, it has been demonstrated that these proteins have evolved through gene translocation followed by gene fusion [10]. Studies of such naturally occurring enzyme aggregates has been extensive and at this time even the three-dimensional structure has been determined for a couple of them [11,12].

Preparation of Artificial Bifunctional Enzymes

A bifunctional enzyme can be prepared by ligating the structural genes of two enzymes. The translational stop signal at the 3′-end of the first gene and the promoter region at the 5′-end of the second gene are removed and, upon ligation, the novel chimeric gene thereby encodes a single polypeptide chain carrying both active sites. Which gene that is selected to be in the *N*-terminus of the final gene product is chosen arbitrarily. Most often, the choice is dependent on the access of restriction enzyme sites and the availability of a three-dimensional structure of the proteins. When ligating structural genes of oligomeric enzymes it can sometimes be confusing to employ notions such as subunits or monomers. Therefore, it can be advantageous to use an alternative nomenclature. Thus, I wish to introduce 'fumer' as a designation for the gene product of two genetically fused structural genes. Upon ligation of two monomeric proteins, the hybrid protein still will have a monomeric character. However, when fusing a dimeric protein with a monomeric, the hybrid protein will most likely maintain a native form of aggregation associated with the oligomeric protein. Thus, the artificial enzyme consists of two associated 'fumers' and must be regarded as a dimeric protein. These and other oligomeric configurations are illustrated in Fig. 2. Over the years, we have prepared a variety of artificial bifunctional enzymes by gene fusion. Various physical and chemical properties have been determined for each of them.

β-Galactosidase/galactose dehydrogenase

The structural gene of β-galactosidase (from *Escherichia coli*), a tetrameric enzyme, was ligated to the 3′-end of galactose dehydrogenase (from *Pseudomonas fluorescens*), a dimeric enzyme [13]. The hybrid enzyme catalyses the sequential hydrolysis of lactose followed by the oxidation of the galactose formed to the corresponding lactone:

$$\text{lactose} \xrightarrow{\beta\text{-galactosidase}} \begin{matrix}\text{galactose}\\ +\ \text{glucose}\end{matrix} \xrightarrow[+\ \text{NAD}]{\text{galactose dehydrogenase}} \begin{matrix}\text{galactonolactone}\\ +\ \text{NADH}\end{matrix}$$

Both enzymes retained a high degree of their native specific activities in spite of the fusion. When corrected for the increase in molecular mass caused by the fusion, these specific activities correspond to 90–100% and 40–50% for native β-galactosidase and galactose dehydrogenase, respectively. A number of physical and chemical effects caused by the fusion have been studied, such as subunit interactions and stability. When ligating the structural genes of two oligomeric enzymes, the formation of an extensive protein aggregation (a

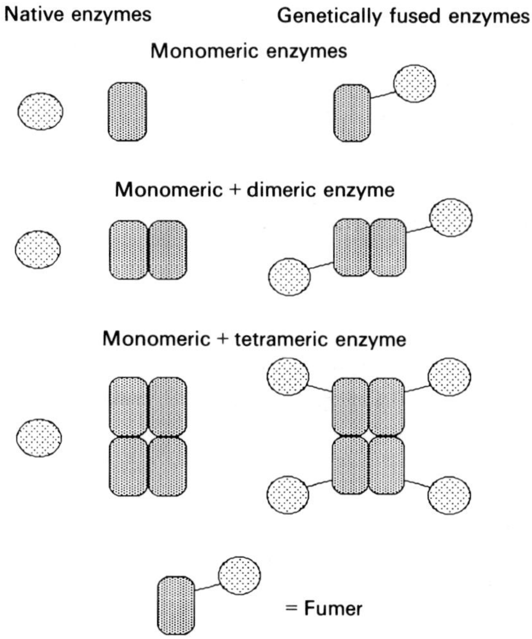

Fig. 2. *Schematic illustration of subunit interactions in genetically fused oligomeric enzymes*

polymer) could be expected due to the interactions between protein 'fumers' (Fig. 3). However, no enzymic activity was found to be associated with insoluble particles. The bifunctional enzyme exhibited mainly a tetrameric (1/3) and a hexameric (2/3) aggregation. This form of 'fumer' interaction might be interpreted as the partial formation of higher aggregates since native β-galactosidase is predominantly found as tetramers. Owing to the dimeric nature of native galactose dehydrogenase it seems plausible that these moieties are able to interact with each other within this hybrid enzyme unit. On closer examination of the fusion protein, it was observed that the tetramer exhibited a higher relative galactose dehydrogenase activity than the hexamer. This might be interpreted as if the extra galactose dehydrogenase parts are not able to interact with each other in the hexamer compared with the tetramer and hence are inactive or at least less active. However, at the present stage it cannot be excluded that the galactose dehydrogenase moieties in the hybrid protein are active as monomers.

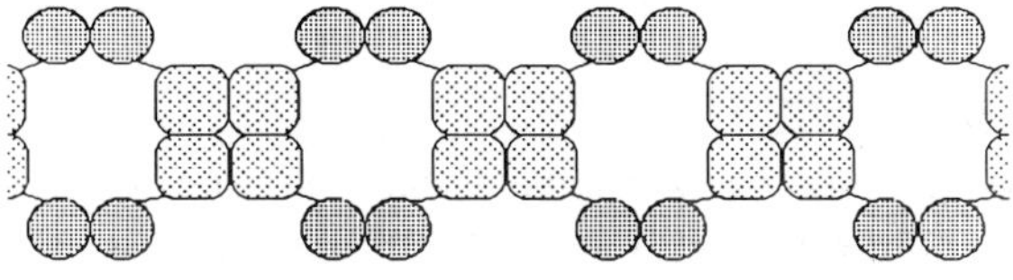

Fig. 3. *Potential formation of a three-dimensional protein network when ligating a dimeric protein with a tetrameric one*

The β-galactosidase part of the bifunctional enzyme proved to be more labile to heat denaturation than the native enzyme. For instance, at 50 °C the hybrid β-galactosidase is quickly denatured, while the native enzyme is unaffected at this temperature. On the other hand, hybrid galactose dehydrogenase was more stable than the native one at the same temperature. Interestingly, such heat incubation caused dissociation of the hybrid enzyme to complexes of dimers. Thus, β-galactosidase is active only in the tetrameric and hexameric forms, while galactose dehydrogenase is active also as dimers. It therefore seems likely that the native subunit interactions of β-galactosidase have been disturbed somewhat causing the observed reduced thermostability.

The proximity effects caused by the gene fusion was evaluated by comparing the kinetic behaviour of the bifunctional enzyme to a similar system with native enzymes. To avoid accumulation of the intermediate galactose, the conditions should be chosen such that the specific activity of the second enzyme, galactose dehydrogenase, is higher than that of β-galactosidase. At pH 7.0, the pH-optimum of β-galactosidase, the ratio of enzyme activities is approximately 1:1. Since galactose dehydrogenase has a much more alkaline pH optimum, pH 10.5, a more favourable situation is achieved by increasing the pH. At pH 8.5 the β-galactosidase activity was thus reduced, while the galactose dehydrogenase activity was increased, giving a ratio of activities of galactose dehydrogenase to β-galactosidase corresponding to 8:1. At this condition, the steady-state rate of NADH formation in the coupled reaction was more than twice that of the native enzymes. Furthermore, the transient times were markedly reduced. At lactose concentrations ranging from 0.2 to 20 mM, the transient time was reduced approximately 4-fold [14].

β-Galactosidase/galactose dehydrogenase/galactokinase

As an extension of the β-galactosidase/galactose dehydrogenase and our previously constructed β-galactosidase/galactokinase hybrids [15–17], we have prepared a trifunctional enzyme exhibiting all three activities on the same polypeptide chain. Owing to its high molecular mass, the hybrid protein is very easy to purify from an *E. coli* homogenate. This tripartite enzyme has mainly tetrameric and octameric configurations and shows a comparable stability to the corresponding bifunctional enzymes. It has proved very useful for analyses of lactose, arabinose and galactose, since each separate enzyme activity easily can be turned 'on and off' by regulating the composition of metal ions in the reaction buffer.

Galactose dehydrogenase/luciferase

To demonstrate further the applicability of gene fusion in the design of bifunctional enzymes, we constructed an artificial hybrid enzyme consisting of galactose dehydrogenase genetically fused to bacterial luciferase (from *Vibrio harveyi*) [18]. This bi-enzyme catalyses reactions (1) and (3) in a reaction sequence catalysed by three enzymes. Thus diaphorase has to be added separately:

$$\text{galactose} + \text{NAD} \xrightarrow{\text{galactose dehydrogenase}} \text{galactonolactone} + \text{NADH} \quad (1)$$

$$\text{NADH} + \text{FMN} \xrightarrow{\text{diaphorase}} \text{NAD} + \text{FMNH}_2 \quad (2)$$

$$\text{FMNH}_2 + \text{O}_2 + \text{RCHO} \xrightarrow{\text{luciferase}} \text{FMN} + \text{H}_2\text{O} + \text{RCOOH} + h\nu \quad (3)$$

It is worth observing that the natural occurrence of such bifunctional enzymes *in vivo*, which catalyse consecutive reactions not immediately adjacent, is not uncommon. The earliest reported example of a multienzyme polypeptide is aspartate kinase I/homoserine dehydrogenase I from *E. coli* where each polypeptide chain has two active sites catalysing the following reactions [19]:

$$\text{ATP} + \text{L-aspartate} \rightarrow \text{ADP} + \text{4-phospho-L-aspartate}$$

$$\text{L-homoserine} + \text{NAD}^+ \rightarrow \text{L-aspartate-}\beta\text{-semialdehyde} + \text{NADH}$$

These two reactions are the first and third steps in the biosynthesis of leucine, threonine and lysine from aspartate.

The practical use of bacterial luciferases has been thoroughly examined by several investigators. Much work has been undertaken to improve the stability and analytical performance of the enzyme. Notably, co-immobilization of bacterial luciferase with various dehydrogenases and diaphorase has, due to close proximity between the enzymes, generated simple and sensitive systems for NAD and metabolite quantifications [20]. However, luciferase has a high tendency to be deactivated by various chemical reagents used for immobilization, but drawbacks associated with such chemical coupling procedures can be circumvented by ligating the protein of interest to bacterial luciferase at the DNA instead of the protein level.

Bacterial luciferase is a heterodimeric enzyme with two different subunits, α and β encoded by the *lux*A and *lux*B genes. The single active centre of luciferase is located on the α-subunit. The role of the β-subunit is unknown, but it is necessary for the enzymic activity of luciferase. Therefore, we decided to employ this latter subunit for fusion purposes. The 5′-terminal end of *lux*B was thus ligated to the 3′-end of the galactose dehydrogenase gene.

The specific activity for the purified fusion protein decreased only slightly in the gene fusion. We found that at least 75% of the activity was left of the luciferase part of the conjugate and that about 90% of the activity was retained of galactose dehydrogenase compared to the native enzymes. Fig. 4 gives a schematic representation of the luciferase/galactose dehydrogenase conjugate.

By analogy to the β-galactosidase/galactose dehydrogenase fusion protein, this hybrid galactose dehydrogenase proved to be more thermostable than the native enzyme. Surprisingly, the reverse situation occurred when the enzymes were treated with urea. The galactose dehydrogenase part of the fusion protein lost most of its activity in the presence of 4 M-urea, while the native enzyme retained its activity at up to 6 M-urea. Hybrid luciferase showed no such differ-

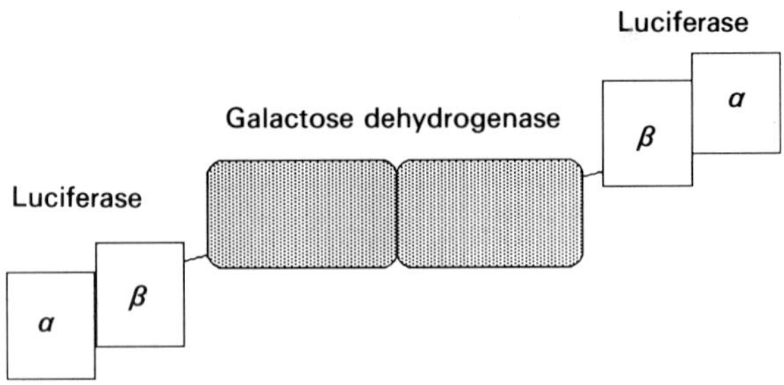

Fig. 4. *Configuration of the galactose dehydrogenase/luciferase conjugate*

ences compared with native luciferase, implying that the structure of luciferase has not been modified. This was also supported by the fact that the emitted wavelength profile from the recombinant enzyme was identical to that of the native enzyme giving an almost unambiguous indication of no structural modifications.

The conjugate was utilized to monitor the bioluminescence at different concentrations of galactose and NAD employing an excess of diaphorase. The conjugate proved to generate a higher bioluminescent peak value than an identical system with native enzymes. This phenomenon was obtained using galactose concentrations over a broad range, normally giving at least a 2-fold increase in bioluminescence. There are several plausible explanations for this apparently higher efficiency for the conjugate, such as: (1) association of diaphorase to the hybrid enzyme giving a three-enzyme complex showing proximity effects or (2) modified enzyme characteristics for the conjugate due to the fusion of the enzymes. Protein–protein interactions are difficult to verify experimentally. The existence of strong intermolecular association between the diaphorase and the recombinant enzyme can be excluded since no such interactions have been detectable during gel-filtration experiments.

In the described examples it is hardly realistic to assume that the active centres should be favourably oriented in relation to each other. Other grounds for the observed kinetic behaviour must be found. As proposed by several authors an unstirred layer surrounds the protein molecule [21]. A plausible explanation could thus be the possibility of the intermediate diffusing along the surface of the protein. It does not thereby need to diffuse out to the surrounding bulk medium before the consecutive reaction takes place. No doubt, the kinetic effects can to a certain extent also be explained by modified kinetic properties of the native enzymes caused by the gene fusion. Notably, slight changes in the K_m values have been observed. In consideration of these various causes, we are presently making theoretical simulations of the described systems to clarify the grounds for the reduced transient times and the increased steady-state rates.

Linker region: metabolic engineering

It can easily be envisaged that the preparation of artificial bifunctional enzymes will be most beneficial in directing a substrate to a preferred metabolic route *in vivo*. Owing to proximity effects, the intermediate can be transferred efficiently to the second enzyme instead of a competing enzyme. This was illustrated by preparing a hybrid between β-galactosidase and galactokinase. These are the first two enzymes in the pathway of lactose metabolism of *E. coli*. Lactose is initially hydrolysed to glucose and galactose which in turn is phosphorylated to galactose 1-phosphate:

$$\text{lactose} \xrightarrow{\beta\text{-galactosidase}} \begin{array}{c}\text{galactose}\\ +\ \text{glucose}\end{array} \xrightarrow[+\ \text{ATP}]{\text{galactokinase}} \text{galactose 1-phosphate}$$

To evaluate further the metabolic effects caused by an artificial bifunctional enzyme a series of similar β-galactosidase/galactokinase hybrids were prepared with only small differences in the linker regions according to the following scheme:

Strain	
105:	Gln-Gly-Ile- Arg-Arg-Pro-Ala-Ala-Lys-Leu-Ile- Asp-Asp-Lys-Leu-Ser-Asn-Met-Ser-Gln
205:	Gly-Asp-Pro-Ser- Thr-Cys-Ser- Gln-Phe-Gln
205pro:	Gly-Asp-Pro-Pro-Pro-Pro-Pro-Ser- Thr-Ser- Gln-Phe-Gln
205gly:	Gly-Asp-Leu-Glu-Gly-Gly-Gly-Gly-Ala-Ser- Thr-Ser- Gln-Phe-Gln

Accordingly, the linkers have different lengths as well as various physical properties such that the proline and glycine segments should give the connecting region a rigid and a flexible character, respectively. When these hybrid enzymes were expressed in *E. coli* grown on minimal medium with lactose as carbon source, however, no difference in growth rates for the separate strains could be observed. Therefore, to demonstrate any differences between the β-galactosidase/galactokinase hybrids, a plasmid expressing galactose dehydrogenase was introduced together with the bifunctional enzyme. The reason for this is that galactose dehydrogenase can compete with galactokinase for the intermediate galactose formed by the β-galactosidase:

$$\text{lactose} \xrightarrow{\beta\text{-galactosidase}} \text{galactose} \begin{array}{l}\nearrow \text{galactose 1-phosphate}\\ \searrow \text{galactonolactone}\end{array}$$

The galactonolactone represents a metabolic end-product in *E. coli*. Galactose dehydrogenase hence operates as a scavenger enzyme. When grown on lactose a pronounced difference in growth rates between the individual strains was observed. Strain 205, having the shortest linker, grows markedly faster than strain 105, whereas 250pro and 205gly have growth rates in between.

Alternative preparation of multifunctional enzymes

When designing more complex enzyme aggregates consisting partly of oligomeric enzymes, it can be advisable (in respect of time and labour) to take advantage of these subunit interactions to assemble protein aggregates *in vivo*. For instance, when preparing a complex consisting of three enzymes, A, B and

C, where C is an oligomeric enzyme, a three-enzyme conjugate can be obtained by preparing the fusion proteins A/C and B/C and expressing them together *in vivo*. A fraction of the total protein will then consist of an A/C–C/B complex. Using this technique we have, for instance, prepared a β-galactosidase/galactose dehydrogenase/luciferase conjugate by expressing a β-galactosidase/galactose dehydrogenase and a galactose dehydrogenase/luciferase plasmid together in the same cell. With the aid of the galactose dehydrogenase subunit interactions the trifunctional conjugate can then be formed. Approximately 10–20% of the total galactose dehydrogenase activity was found in this complex, which can easily be isolated by gel filtration.

Discussion: Prospects

From a practical point of view, the utilization of hybrid bifunctional enzymes in enzyme technology offers many advantages. Their purification is simplified since only the affinity of one part of the protein is required during a chromatographic procedure. Such ready-made enzyme sequences should facilitate their application in analysis using a variety of biosensors. For instance, the β-galactosidase/galactose dehydrogenase hybrid is most useful for lactose analyses from various biological samples. Under such circumstances it is very beneficial to take advantage of the inherent substrate channelling associated with bifunctional enzymes. For instance, galactose dehydrogenase also accepts L-arabinose as substrate. Arabinose therefore interferes in lactose analyses using separate native enzymes. Owing to substrate channelling such interference can often be almost totally reduced, since galactose dehydrogenase preferentially uses the galactose formed from β-galactosidase. The improved sensitivity of the bifunctional enzymes is of course another very attractive property in analytical systems. In the case of galactose dehydrogenase/luciferase, the sensitivity of the system could thus be improved at least twice.

In this paper, I have focused all my attention on the design of bifunctional proteins consisting of two entire native enzymes. The described techniques should, however, also be applicable to gene fusion of only parts of enzymes. The tertiary structure of proteins suggests that the polypeptide chain can fold into distinct compact regions. These domains have often been considered as functional units of a protein. The concept of domains also as a genetic unit has been underlined by several authors as indicated by the correlation between exons and domains of a protein [22], which should allow their isolation with conventional molecular biological methods. For instance, FAD is covalently bound to 6-hydroxy-D-nicotine oxidase (6-HDNO) [23]. The DNA fragment encoding the 'signal sequence' for flavinylation of this protein can be readily isolated and fused to an alternative structural gene. Thus, the FAD-binding portion of 6-HDNO was isolated and fused to the structural gene of β-galactosidase. The FAD moiety could then be used to couple apo glucose oxidase to the hybrid enzyme and thereby be utilized to generate an alternative bifunctional enzyme (L. Bülow, unpublished work). Even smaller structural elements occurring in proteins can either be isolated or obtained through chemical synthesis of the corresponding DNA fragment. By making combinations in this naturally occurring repertoire of 'building blocks', the binding or catalytic sites of a protein can be modified. This method might even be used to create entirely novel enzymes.

Helén Carlsson, Christer Lindbladh, Peter Ljungcrantz, Klaus Mosbach, Mats-Olle Månsson and Mats Persson are gratefully acknowledged for their contributions.

References

1. Welch, G. R. (ed.) (1985) *Organized Multienzyme Systems*, Academic Press, N.Y.
2. Gaertner, F. H. (1978) in *Microenvironmental and Metabolic Compartmentation* (Srere, P. & Easterbrook, R. W., eds.), pp. 345–352, Academic Press, New York
3. Kula, M.-R. & Wandrey, C. (1988) *Methods Enzymol.* **136**, 9–21
4. Scheller, F., Siegbahn, N., Danielsson, B. & Mosbach, K. (1985) *Anal. Chem.* **57**, 1740–1743
5. Mosbach, K. & Mattiasson, B. (1970) *Acta Chem. Scand.* **24**, 2093–2100
6. Shimutzu, S. Y. & Lenhoff, H. M. (1979) *J. Solid Phase Biochem.* **4**, 75–122
7. Srere, P. A., Mattiasson, B. & Mosbach, K. (1973) *Proc. Natl. Acad. Sci. U.S.A.* **70**, 2534–2538
8. Koch-Schmidt, A.-C., Mattiasson, B. & Mosbach, K. (1977) *Eur. J. Biochem.* **81**, 71–78
9. Månsson, M.-O., Siegbahn, N. & Mosbach, K. (1983) *Proc. Natl. Acad. Sci. U.S.A.* **80**, 1487–1491
10. Kirschner, K. & Bisswanger, H. (1976) *Annu. Rev. Biochem.* **45**, 143–166
11. Priestle, J. P., Grütter, M. G., White, J. L., Vincent, M. G., Kania, M., Wilson, E., Jardetssy, T. S., Kirschner, K. & Jansonius, J. N. (1987) *Proc. Natl. Acad. Sci. U.S.A.* **84**, 5690–5694
12. Hyde, C. C., Ahmed, S. A., Padlan, E. A., Miles, E. W. & Davies, D. R. (1988) *J. Biol. Chem.* **263**, 17857–17871
13. Buckel, P. & Zehelein, E. (1981) *Gene* **16**, 149–159
14. Ljungcrantz, P., Carlsson, H., Månsson, M.-O., Buckel, P., Mosbach, K. & Bülow, L. (1989) *Biochemistry* **28**, 8786–8792
15. Bülow, L., Ljungcrantz, P. & Mosbach, K. (1985) *Bio/Technology* **3**, 821–823
16. Bülow, L. & Mosbach, K. (1986) *Proc. Enzyme Engineering* I, Cambridge, U.K., IBC Techn. Services Ltd.
17. Bülow, L. (1987) *Eur. J. Biochem.* **163**, 443–448
18. Waddle, J. J., Johnston, T. C. & Baldwin, T. O. (1987) *Biochemistry* **26**, 4917–4921
19. Umbarger, H. E. (1978) *Annu. Rev. Biochem.* **47**, 533–606
20. Weinhausen, G. & DeLuca, M. (1982) *Anal. Biochem.* **127**, 380–388
21. Hagler, A. & Moult, J. (1978) *Nature (London)* **272**, 222–226
22. Gilbert, W. (1986) *Cell (Cambridge, Mass.)* **46**, 151–154
23. Brandsch, R. & Bichler, V. (1987) *FEBS Lett.* **224**, 121–124

Biochem. Soc. Symp. **57**, 135–141
Printed in Great Britain

Restructuring Catalysis in the Mandelate Pathway

DAVID C. NEIDHART*, P. LYNNE HOWELL*, GREGORY A. PETSKO*, JOHN A. GERLT†, JOHN W. KOZARICH†, VINCENT M. POWERS‡ and GEORGE L. KENYON‡

* *Department of Chemistry, Massachusetts Institute of Technology, Cambridge, MA 02139, U.S.A.,* † *Department of Chemistry and Biochemistry, University of Maryland, College Park, Maryland 20742, U.S.A. and* ‡ *Department of Pharmaceutical Chemistry, University of California, San Francisco, California 94143, U.S.A.*

Synopsis

Mandelate racemase (MR) is the first enzyme in the bacterial pathway that converts mandelic acid to benzoic acid. The mandelate pathway can utilize either enantiomer of mandelate because this enzyme interconverts them. We have solved the structure of MR at 2.5 Å resolution. The enzyme is almost identical in conformation to another bacterial enzyme, muconate lactonizing enzyme (MLE). Both enzymes are TIM-barrel proteins. This result has profound implications for the evolution of enzymic function and the origin of metabolic pathways. It also implies that it should be possible to transform one enzyme into the other by site-directed mutagenesis.

The Mandelate Pathway

The evolution of catalytic function is a sufficiently intriguing problem that an entire volume of recent research reports has been devoted to it [1]. Some time ago, we became convinced that it was possible to investigate this problem by the combined methods of X-ray crystallography, molecular biology and mechanistic enzymology. Such an investigation entails the choice of a suitable pathway for investigation as the first step. The criteria for the choice are easy to define: (1) the pathway should be of relatively recent evolutionary origin, since a primitive pathway might have too many ancestral enzymes in it and since ongoing mutations may obscure the origins of older pathways; (2) the pathway should be small so that the number of enzymes to investigate is manageable; (3) the pathway should contain a broad range of enzymic reaction types to avoid the trivial result that one enzyme could simply be replicated repeatedly to produce a pathway with similar functionalities; (4) the pathway should be microbial so that the power of genetics can be readily exploited to dissect structure–function relationships and selection pressures on the pathway; and (5) the enzymes of the pathway should be organized into an operon so that it would be straightforward to clone and sequence all of them.

These considerations rule out many of the most familiar pathways such as those for glycolysis and the tricarboxylic acid cycle. But there is an obvious

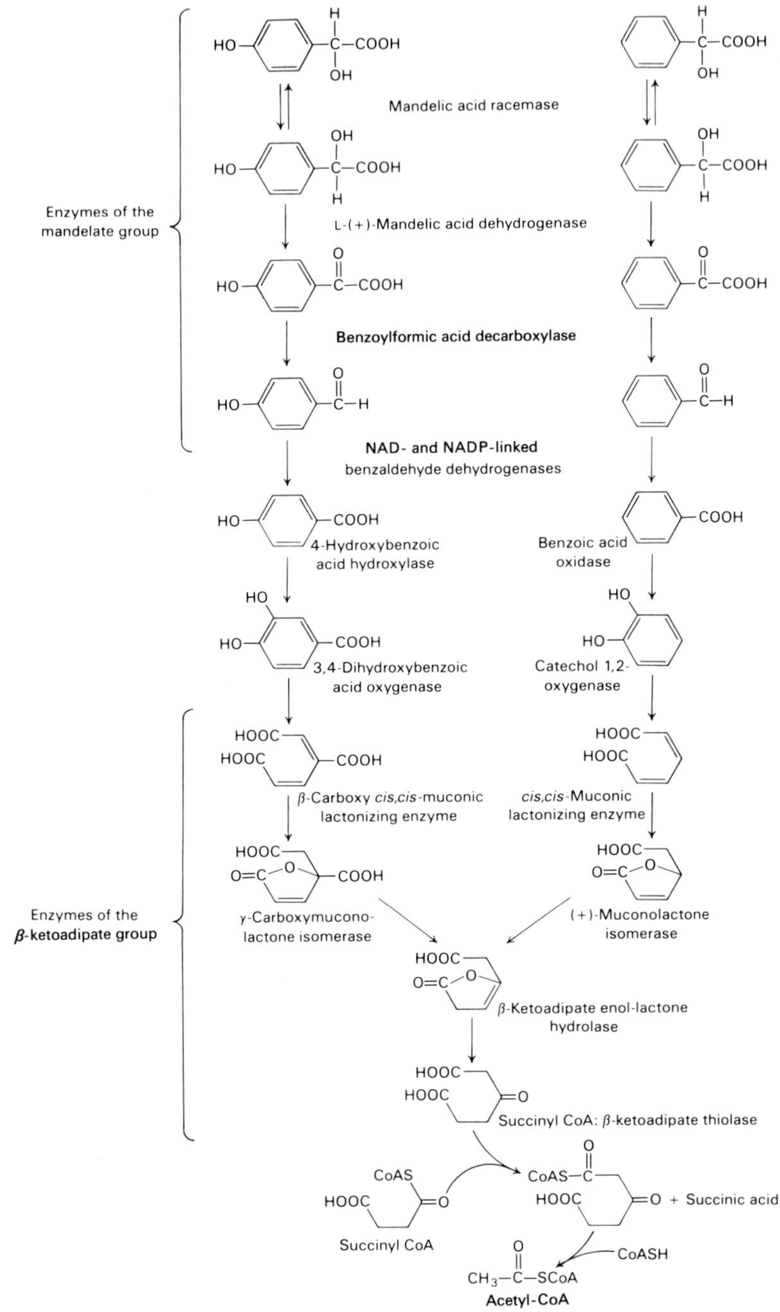

Fig. 1. *The reactions of the mandelate pathway in Pseudomonas putida and the connection to the β-ketoadipate pathway*

Redrawn from Halpin *et al.* (1981) *Biochemistry* **20**, 1525–1533.

candidate, a pathway that in fact was the first one to be identified as having an operon organization for its genes: it is found in only a few pseudomonads, so it is likely that it is of recent evolutionary origin; it appears to have only five enzymes, arranged in a simple fashion, but covering a broad range of chemical reaction types — it is called the mandelate pathway (Fig. 1).

Micro-organisms that contain the enzymes of the mandelate pathway are able to use either optical isomer of mandelic acid as a sole source of carbon and energy. They do this by converting mandelate into benzoic acid and then to catechol, which feeds into the β-ketoadipate pathway, eventually producing both acetyl-CoA and succinate (Fig. 1). Only the *S*-isomer of mandelate is competent to be oxidatively degraded in the pseudomonads studied in this work, but the organism can use the *R*-isomer as well. The enzyme that enables it to do so is MR, which is apparently the first enzyme of the mandelate pathway.

Mandelate Racemase

MR catalyses the interconversion of *R*- and *S*-mandelic acids. It appears to be an octameric enzyme of overall molecular mass 320 000 Da and is now known to be composed of eight identical polypeptide chains. There is no evidence that the enzyme is either co-operatively or allosterically regulated. A divalent metal ion is required for catalytic activity; the physiological metal ion seems to be magnesium, but manganese is able to substitute, albeit with somewhat reduced activity [2].

MR belongs to the class of enzymes, known as racemases, that catalyse the interconversion of optical isomers. It is a very interesting class. First of all, these enzymes must be able to bind and utilize substrates of opposite chirality. This is completely contrary to the behaviour of other enzymes, which are normally absolutely specific for a single optical isomer of the molecules on which they operate. Secondly, the class of enzymes is striking in its diversity: some racemases require a divalent metal ion for activity, while others need no metal, but use a cofactor such as pyridoxal phosphate. Finally, since racemases are single substrate/single product enzymes, it is possible to examine the structures of their enzyme–substrate complexes directly by straightforward protein crystallographic techniques [3]. We were interested to learn how MR employs magnesium ions and binds both *R*- and *S*-mandelate with equal affinity, as well as how it became part of the mandelate pathway in the first place. Towards this end, we undertook the crystallization and three-dimensional structure determination of MR and of its Michaelis complex with mandelate by the standard methods of protein crystallography.

The Crystal Structure of Mandelate Racemase

We first grew large single crystals of MR under several different conditions. In buffered polyethyleneglycol at pH 8, two distinct tetragonal crystal forms can be produced. One form is shaped as rectangular plates having the symmetry of space-group I422, with unit cell dimensions $a = b = 153$ Å, $c = 181$ Å. There are two protomers, each of molecular mass 38 438 Da in the crystallographic asymmetric unit. A second form, which can coexist with the first, is octahedrally shaped. These crystals have the symmetry of space-group I4, with unit cell dimensions $a = b = 113$ Å, $c = 124$ Å. Again, the asymmetric unit appears to consist of two-MR protomers. Both of these crystal forms demonstrate diffraction to beyond 3 Å resolution [4].

Ammonium sulphate solutions produced a third crystal form. These crystals are tetragonal-shaped prisms. They grow quite slowly, but can be grown to very large size: one specimen of almost 2 mm on a side has been grown and 0.5 mm-sized crystals are typical. The diffraction pattern from these crystals extends well beyond 2.5 Å resolution and shows moderate sensitivity to irradiation. The crystals have the symmetry of space-group I422, the same as that of the rectangular plates grown from polyethyleneglycol, but this time the unit cell dimensions are $a = b = 125$ Å, $c = 105$ Å. There is only a single protomer of MR in the asymmetric unit [5].

The small size of the asymmetric unit of the ammonium-sulphate-grown crystals, which we refer to as Form III, made this crystal form ideal for structure determination of this protein. We proceeded to collect a complete set of native X-ray data to 2.5 Å resolution on a single-counter diffractometer equipped with a conventional sealed-tube X-ray source. Our heavy-atom derivative search was simplified by the fact that MR requires a divalent metal ion for catalytic activity, and europium is known to substitute for divalent metals in many metalloenzymes [6]. We were able to prepare an excellent single-site derivative by replacing the native magnesium ion with europium, and quickly found two other derivatives as well: a mercury derivative that bound to the two cysteine residues known to be present in each protomer, and a platinum derivative. We calculated a 3 Å electron density map using phases from all three derivatives and immediately could distinguish the molecular boundary from the surrounding solvent. We used solvent flattening [7] to improve the phases and make a new map. It also proved to be very easy to trace the course of the polypeptide chain in this map. When we did, we were intrigued to find that the enzyme belongs to a recognizable structural class: the so-called $\alpha\beta$-barrel proteins exemplified by the first member of the family, triose phosphate isomerase (TIM). About one-half of the structure of MR is in the form of a TIM-barrel. The remainder of the protein is organized into two additional folding domains (Fig. 2). The three-domain monomer units are packed in two square-planar arrangements to yield the octameric protein, which overall has the same symmetry (422) as the space-group of this crystal form. This is illustrated in Fig. 3.

We were even more intrigued when we displayed the overall structure on the molecular graphics system. It looked familiar. In fact, it looked just like the polypeptide chain fold of MLE, an enzyme of the β-ketoadipate pathway the three-dimensional structure of which was solved by Goldman *et al.* some years ago [8]. The β-ketoadipate pathway is the main pathway into which the product of the mandelate pathway feeds, so this structural similarity prompted us to look for evolutionary homology between these two proteins. Their

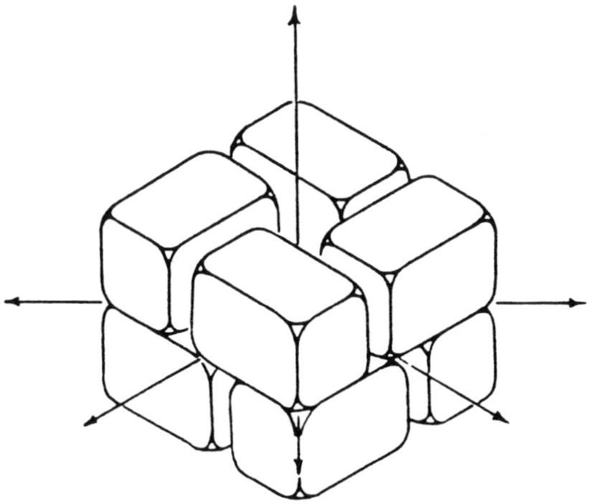

Fig. 2. *The symmetry arrangement of the eight subunits of mandelate racemase*

sequences had never been compared, so we performed that comparison using unpublished data for the MLE sequence and the MR sequence obtained by one of us (J.A.G.). The sequences turned out to be about 30% identical, indicating that MR and MLE do indeed share a common ancestor.

It is tempting to conclude that this ancestor is MLE itself. When the mandelate pathway evolved, and an enzyme was needed that catalysed the racemization of mandelic acid, the gene coding for MLE was duplicated and the duplicate gene then mutated independently to form MR. Support for this view comes from a consideration of the reactions catalysed by the two

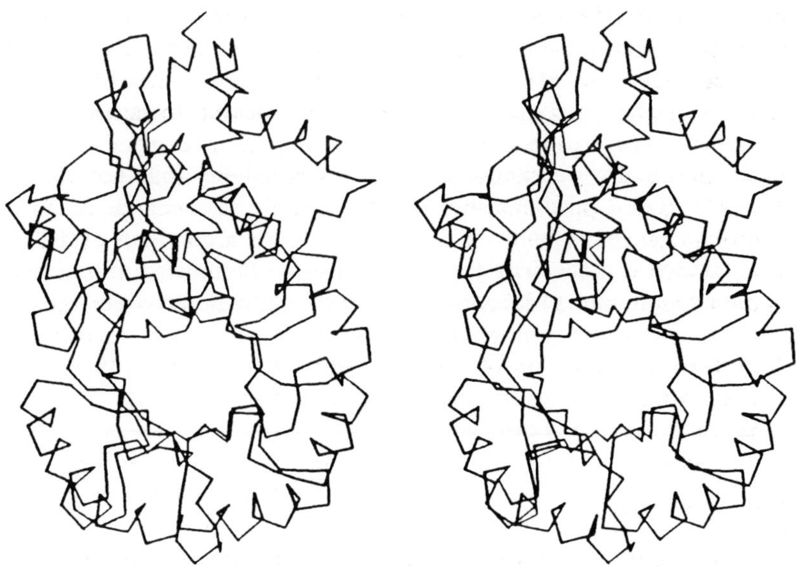

Fig. 3. *Stereo drawing of the folding of the backbone of a single subunit of mandelate racemase*
The drawing for muconate lactonizing enzyme would be essentially identical.

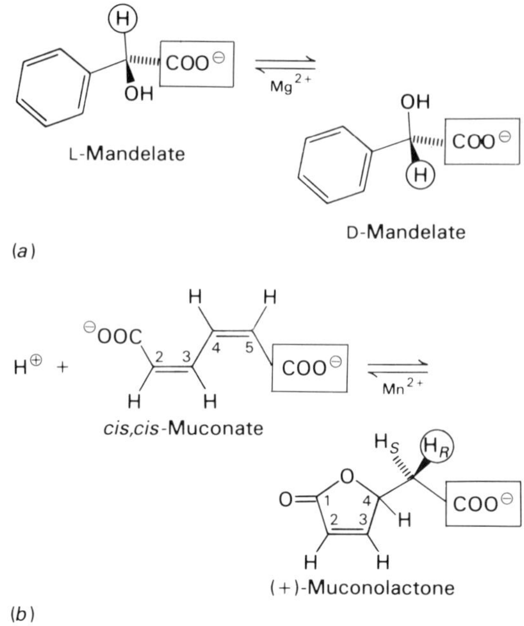

Fig. 4. *The reactions catalysed by* (*a*) *mandelate racemase and* (*b*) *muconate lactonizing enzyme*
Note that the chemistry of the muconate lactonizing enzyme reaction in the reverse direction is similar (α-proton abstraction to a carboxyl) to that for mandelate racemase.

enzymes: although superficially they seem quite different, both enzymes actually catalyse the abstraction of an α-proton to a carboxylate, although MR does this in both directions and MLE does this only in its reverse direction (Fig. 4). Moreover, both enzymes use a divalent metal ion, and the ligands to the metal are among the residues that are conserved when the sequences are aligned. We believe that MR and MLE are striking and compelling examples for the evolution of catalytic function and new metabolic pathways by the recruitment of enzymic properties from pre-existing pathways.

Of course, formally we cannot exclude the possibility that both MR and MLE evolved from a third enzyme, one that has not yet been identified. That would not change the arguments presented here for a common evolutionary origin, nor would it necessarily prevent our using site-directed mutagenesis to imitate the evolutionary process by which these catalytic activities arose. The tools of molecular biology now allow us to carry out a plan to recapitulate and learn from the evolution of catalytic function in MR and MLE and, we expect, from other enzymes in the mandelate pathway as well. We call this process the restructuring of catalysis.

Restructuring Catalysis

It is possible that we can use site-directed mutagenesis to convert one of these enzymes into the other. The ligands to the metal are conserved, and each enzyme can use the other's preferred metal ion already, so that site need not be touched. The remainder of the problem is straightforward: to re-engineer the specificity pocket to take the other substrate and to provide the necessary catalytic residues. Preliminary inspection of the MLE active site indicates that it has one of the two bases that we see in the MR active site, so we may only need to delete a single residue in the MR active site to take care of the chemistry problem. These experiments are in progress. Of course, we do not just have to do this restructuring ourselves: the bacteria did it under selection pressure, and we may be able to imitate that process in the laboratory by subjecting an organism that has, say, only the MLE gene (but in many copies on a plasmid) and no MR gene, to the requirement for growth on mandelate as a sole source of carbon and energy. That may be too severe. If it is, we will have to design milder selection strategies and approach the problem more gradually. We are trying that, too. Whatever the outcome, we should have fun trying! And even if we fail, we believe that this study has already provided a valuable glimpse into the evolutionary origins of a metabolic pathway. And we still have four more enzymes to go!

We thank Dr Adrian Goldman for MLE coordinates and many useful comments. Dr Gregory K. Farber provided many valuable discussions about the structural relationships in TIM-barrel enzymes. This work supported by USPHS grants AR17323 (to G.L.K.) and GM40570 (to J.A.G., G.L.K., J.W.K. and G.A.P.).

References

1. *Cold Spring Harbor Symp. Quant. Biol.* (1987) **LII**
2. Whitman, C. P., Hegeman, G. D., Cleland, W. W. & Kenyon, G. L. (1985) *Biochemistry* **24**, 3936–3942
3. Alber, T., Banner, D. W., Bloomer, A. C., Petsko, G. A., Phillips, D. C., Rivers, P. S. & Wilson, I. A. (1981) *Philos. Trans. R. Soc. London B* **293**, 159–171
4. Neidhart, D. J., Powers, V. M., Kenyon, G. L., Tsou, A. Y., Ransom, S. C., Gerlt, John A. & Petsko, G. A. (1988) *J. Biol. Chem.* **263**, 9268–9270
5. Neidhart, D. J. (1988) Ph.D. thesis, Massachusetts Institute of Technology, U.S.A.
6. Farber, G. K., Machin, P., Almo, S. C., Petsko, G. A. & Hajdu, J. (1988) *Proc. Natl. Acad. Sci. U.S.A.* **85**, 112–115
7. Wang, B. C. (1985) *Methods Enzymol.* **115**, 90–112
8. Goldman, A., Ollis, D. L. & Steitz, T. A. (1987) *J. Mol. Biol.* **194**, 143–162

Biochem. Soc. Symp. **57**, 143–145
Printed in Great Britain

Identification and Design of Binding Determinants in Proteins

JAMES A. WELLS and BRIAN C. CUNNINGHAM

Department of Biomolecular Chemistry, Genentech, Inc., 460 Pt. San Bruno Blvd, South San Francisco, CA 94080, U.S.A.

Synopsis

Mutagenic analyses are described for probing and designing binding determinants of two different proteins. The first protein is an enzyme, subtilisin, for which structural models were available for the enzyme–substrate complex. The second protein is human growth hormone for which a structural model of its complex with receptor was not available. In both cases, it was possible to identify and engineer binding determinants that lead to large changes in affinity.

Introduction

Homologous proteins can vary widely in functional properties, especially binding specificity. This is perhaps most evident in the family of mammalian serine proteinases where homologous enzymes such as trypsin, elastase and chymotrypsin exhibit dramatically different substrate specificities (for a review see [1]). Comparisons of their X-ray elucidated structures revealed important structural differences between these enzymes in the residues that contact the substrate.

Results and Discussion

To test the importance of residues making direct contact with the substrate, we have engineered the substrate-binding site of a serine proteinase, subtilisin, that is convergently related to the mammalian family of enzymes (for a review see [2]). Mutations within the substrate-binding site that create changes in electrostatic [3], steric and hydrophobic interactions [4], cause large (and somewhat predictable) changes in substrate specificity.

It is also possible to recruit the substrate specificity profile from one natural variant subtilisin (from *Bacillus amyloliquefaciens*) into another (*Bacillus licheniformis*) by changing only residues that can make direct contact with the substrate [5]. These studies show that engineering contact residues identified by high-resolution structural analysis can have huge effects on substrate specificity.

It is often, in fact, usually the case, that for a protein of interest one does not have a high-resolution structure of the protein–ligand complex. This was precisely our situation when faced with the problem of locating the receptor-binding determinants in human growth hormone (hGH). This was important in order to determine the feasibility of designing a small molecule analogue of this hormone.

Two complementary mutagenic strategies have been employed to define residues in hGH that are most important for binding to the cloned hGH receptor from liver. In the first approach, termed homologue-scanning mutagenesis [6], segments (7–30 residues long) of non-binding homologues (human prolactin or placental lactogen) were substituted throughout the hGH molecules (191 residues in length). Receptor binding was dramatically disrupted by a small set of these segment-substituted molecules that mapped primarily to two separate regions of the hGH sequence; these regions formed a patch when mapped on to a low-resolution model derived from porcine growth hormone [9].

A higher-resolution strategy, termed alanine-scanning mutagenesis, was then used to identify specific side-chains within the binding patch that were most important in binding. From 62 single alanine mutants, 12 side-chains were

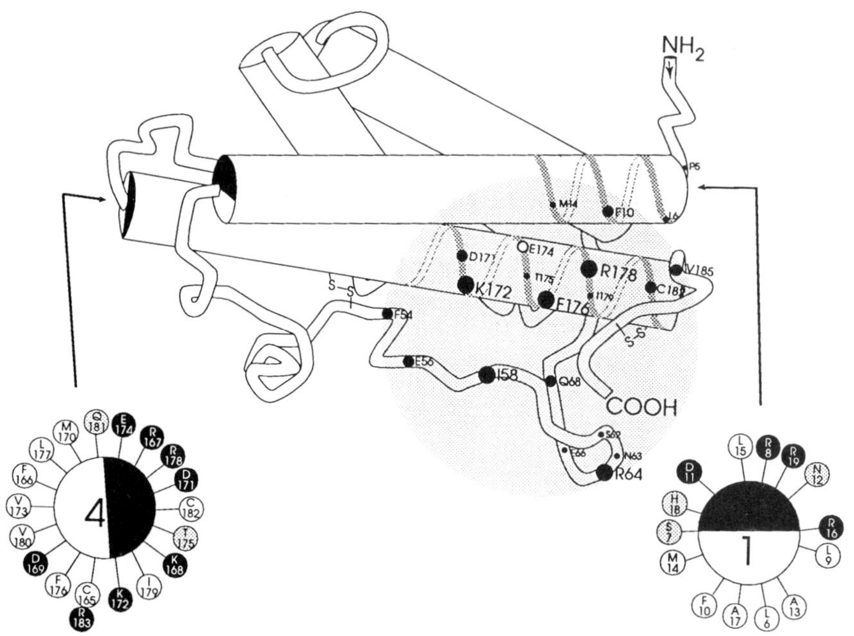

Fig. 1. *Location of residues in hGH that strongly modulate its binding to the hGH-binding protein*
Alanine substitutions (serine or asparagine in the case of T175 or R178, respectively) are indicated that cause >10-fold reduction (●), a 4–10-fold reduction (•) or increase (○), or a 2–4-fold reduction (·), in binding affinity [7]. Helical wheel projections in regions of α-helix reveal their amphipathic quality. Note that in helix 4 the most important determinants are on its hydrophilic face (shaded). The structural model [6] was derived from a folding diagram of porcine GH determined crystallographically [9]. The Figure is taken from reference [8] with permission. Copyright 1990 by the AAAS.

identified that caused large reductions in binding affinity (4–20-fold); one mutant, E174A, was found that caused more than a 4-fold increase in affinity. These two systematic approaches produced a high-resolution functional map of residues important for receptor (Fig. 1) or antibody interactions [6,7]. Recently, this information has been used to incorporate the receptor-binding determinants into human prolactin to recruit it to bind to the hGH receptor [8].

In summary, protein residues in direct contact with a ligand are extremely important in conferring binding affinity. These may be localized by direct structural analysis or inferred from mutagenic analysis. By either method of identification, when such binding determinants from homologue X are introduced into another (homologue Y) which may have a dramatically different ligand specificity, it is possible to recruit the specificity property of homologue X into Y. This is a useful protein design strategy because it allows one to combine the most useful functional properties (such as specificity, catalysis, stability, etc.) contained within the family into one homologue.

Reprinted with permission from *Opportunities in Biotransformations* (L. G. Copping, R. E. Martin, J. A. Pickett, C. Bucke & A. W. Bunch, eds.), published for SCI by Elsevier Science Publishers, London, 1990, pp. 32–25. © SCI 1990.

References

1. Kraut, J. (1977) *Annu. Rev. Biochem.* **46**, 331–358
2. Wells, J. A. & Estell, D. A. (1988) *Trends Biochem. Sci.* **13**, 291–297
3. Wells, J. A., Powers, D. B., Bott, R. R., Graycar, T. P. & Estell, D. A. (1987) *Proc. Natl. Acad. Sci. U.S.A.* **84**, 1219–1223
4. Estell, D. A., Graycar, T. P., Miller, J. V., Powers, D. B., Burnier, J. P., Ng, P. G. & Wells, J. A. (1986) *Science* **233**, 659–663
5. Wells, J. A., Cunningham, B. C., Graycar, T. P. & Estell, D. A. (1987) *Proc. Natl. Acad. Sci. U.S.A.* **84**, 5167–5171
6. Cunningham, B. C., Jhurani, P., Ng, P. & Wells, J. A. (1989) *Science* **243**, 1330–1336
7. Cunningham, B. C. & Wells, J. A. (1989) *Science* **244**, 1081–1085
8. Cunningham, B. C., Henner, D. J. & Wells, J. A. (1990) *Science* **247**, 1461–1465
9. Abdel-Meguid, S. S., Shieh, H.-S., Smith, W. W., Dayringer, H. E., Violand, B. N. & Bentle, L. A. (1987) *Proc. Natl. Acad. Sci. U.S.A.* **84**, 6434–6437

Biochem. Soc. Symp. **57**, 147–155
Printed in Great Britain

Antibody-Combining Sites: Prediction and Design

DAVID S. GREGORY, DAVID STAUNTON, ANDREW C. R. MARTIN*, JANET C. CHEETHAM, JAN PEDERSEN and ANTHONY R. REES†‡

Laboratory of Molecular Biophysics, University of Oxford, The Rex Richards Building, South Parks Road, Oxford OX1 3QU, U.K.

Synopsis

For maximum value, a predicted model of an antibody-combining site should have an accuracy approaching that of an X-ray structure (1.6–2.7 Å). In addition, the method by which the combining site is modelled should make no demands on the user of a sort that require arbitrary or subjective decisions to be made during the process. We have made substantial progress towards this objective and some recent results are reviewed. In addition, we describe how the modelling protocols developed can aid in the design of novel features within the antibody-combining site. The particular design example reported here suggests an approach for the introduction of metal-binding sites to create metallo-antibodies. This type of modification may be useful in the design of immunobiosensors, the induction of catalytic activity or simply as an alternative to metal chelates in the preparation of antibodies for imaging.

Introduction

The wide range of specificities exhibited by antibodies is a function of the sequence and length variability of six hypervariable loops or complementarity-determining regions (CDRs) [1] which form the antigen-combining site. These six CDRs supported on a highly conserved β-sheet framework region [2] constitute the variable region of the antigen-binding fragment (Fab). A knowledge of antibody structure is essential for intelligent design of antibody enzymes [3], tailoring of affinity [4] and CDR replacement strategies [5]. However, sequence information vastly exceeds structural information from X-ray crystallography and, until crystallographic structure determination becomes no less routine than sequencing, modelling of structures is necessary. Since the framework region is conserved, it has proved relatively easy to model, whereas the CDRs, by their very nature [6], present a more challenging problem, since accuracy in their modelling is of paramount importance.

* Present address: National Institute for Medical Research, The Ridgeway, Mill Hill, London, NW7 1AA, U.K.

† Present address: Department of Biochemistry, University of Bath, Claverton Down, Bath BA2 7AY, U.K.

‡ Author to whom correspondence should be addressed.

The approaches taken to modelling the antibody-combining site, so far, fall into two groups: knowledge-based and *ab initio*. Knowledge-based approaches have been used to model a number of antibodies, including GLOOP-1–5 [6], J539 [7], HyHEL-10 [8] and D1.3 [9]. Although the methods differ in their detail, the common feature of all the approaches has been to examine only the known antibody crystal structures and select CDRs from these on the basis of length and/or sequence. Although most methods use simple sequence homology to select model loop conformations, Chothia & Lesk [10] have obtained better results by selecting conformations on the basis of the conservation of 'key' residues that affect loop packing or conformation. However, the chief problem with any such method is the limited size of the knowledge base; while this has been extended to include all protein loops in the broader protein modelling field [11], a general database has not been used previously in antibody modelling.

The second approach has been to use *ab initio* conformational search algorithms to saturate the conformational space available to a loop and select an appropriate structure on the basis of its energy, calculated by using an empirical energy function [12]. Whereas this overcomes the limited size of the knowledge base, it fails to make use of the valuable information available in the structural database and, consequently, is extremely expensive in computer time. Representative are the conformational search methods of Moult & James [13] and Bruccoleri & Karplus [14] and the random conformation dynamics method of Fine *et al.* [15]. To-date, the knowledge-based and *ab initio* approaches have had no more than limited success; they are not routinely able to construct all six CDRs with a high level of accuracy.

A Combined Algorithm

Our own approach has been to combine the two methods to produce a protocol that exploits the structural information in the Brookhaven database [16] (the knowledge-based component) and adds to it the ability to saturate conformational space by conformational searching and energy screening (the *ab initio* component).

The combined algorithm developed has already been described in some detail [17]. It has been used to model the combining sites of the anti-lysozyme antibodies HyHEL-5 [18] and GLOOP-2 [6] according to the following procedure. First, the framework region was taken from the Fab X-ray structure. Secondly, CDR conformations were derived by application of the scheme shown in Fig. 1, where the modelling for each CDR is carried out in the presence of the crystal structure conformations of the other five. Thus, the scheme is essentially a single-CDR method and is useful for modelling loop replacements, insertions, deletions and mutations, etc., or for providing a starting model for interpretation of electron density maps.

Recently, we have extended these studies in two ways. The first development has been to combine the single-CDR modelling scheme with the 'canonical structure' method of Chothia, Lesk and their co-workers [10,19]. In this procedure an appropriate framework region is selected from the database of anti-

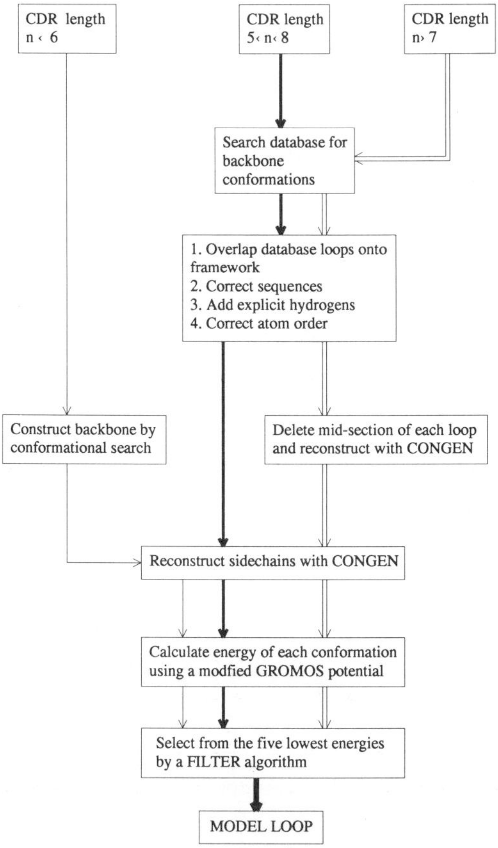

Fig. 1. *Scheme for modelling the three classes of CDR, defined according to length, by the combined algorithm (described in [17])*

The modifications to the GROMOS [23] potential involve removal of the attractive van der Waals and electrostatic terms during evaluation of the energies of each conformer. The FILTER algorithm is described in [24].

body structures based on sequence similarity. This is followed by the modelling of those CDRs for which canonical structures can be identified (usually four of six). Finally, the remaining CDRs are modelled by the combined algorithm. Using this hybrid method, antibody-combining sites can be generated in the absence of any prior crystallographic information about the antibody being modelled, whose all-atom r.m.s. deviations are within the 1–2 Å range (J. C. Cheetham, A. C. R. Martin, C. Chothia, A. Lesk, P. Jeffrey & A. R. Rees, unpublished work). Thus, provided that the backbone conformations of at least four CDRs can be modelled using canonical structures, it should be possible to model any antibody-combining site with confidence.

There are, however, circumstances where canonical structures have not yet been shown to exist for certain CDRs, either because the database is insufficiently large to contain enough representative structures or because certain CDRs are intrinsically more hypervariable in sequence and structure (e.g.

CDR H3). In these situations, the combined method, in the form described above, would differ from antibody to antibody depending on how many CDRs could be confidently placed in canonical groups.

In anticipation of this problem, we began to develop a protocol in which all six CDRs are 'built' on to an empty framework using a development of the algorithm described in Martin *et al.* [17]. Modifications to the algorithm were necessary for the following reasons. Ideally, all six CDRs would be constructed simultaneously. This is impractical in computer time and necessitated an analysis to determine the order in which successive CDRs should be constructed. CDRs constructed early will have fewer constraints on their conformational flexibility than those constructed later. An optimum algorithm has now been achieved and will be described elsewhere. In summary, by using the scheme we have devised, it is now feasible to model the combining site of an antibody by reference only to the amino acid sequences of the variable regions and application of the following protocol: (1) select framework region of each variable domain based on sequence homology; (2) carry out framework sequence substitutions using an automated procedure (J. Pederson & A. R. Rees, unpublished work); (3) orient variable domains based on known packing

Table 1. *Comparison of the complete-combining-site-construction model of GLOOP-2 (unminimized and energy minimized) with the $P2_1$ and P1 crystal structures of GLOOP-2*

The overall values are calculated over the whole structure. Backbone atoms are defined as N, C_α, C, O. (P. D. Jeffrey, G. L. Taylor, R. Griest & A. R. Rees, unpublished work). *B/B*, backbone; EM, energy minimization.

		Global fit r.m.s. deviation			
		$P2_1$		P1	
CDR		Before EM	After EM	Before EM	After EM
L1	All	1.901 Å	1.925 Å	2.092 Å	2.102 Å
	B/B	1.085 Å	1.168 Å	0.954 Å	1.018 Å
	C_α	0.922 Å	0.960 Å	0.862 Å	0.860 Å
L2	All	1.250 Å	1.245 Å	1.101 Å	1.050 Å
	B/B	0.919 Å	0.912 Å	0.739 Å	0.732 Å
	C_α	0.682 Å	0.612 Å	0.662 Å	0.725 Å
L3	All	2.080 Å	1.837 Å	1.996 Å	1.776 Å
	B/B	1.012 Å	0.997 Å	0.848 Å	0.875 Å
	C_α	0.756 Å	0.672 Å	0.754 Å	0.750 Å
H1	All	2.056 Å	2.114 Å	2.037 Å	2.068 Å
	B/B	1.356 Å	1.361 Å	1.424 Å	1.412 Å
	C_α	0.793 Å	0.702 Å	1.029 Å	0.946 Å
H2	All	2.505 Å	2.524 Å	2.227 Å	2.159 Å
	B/B	1.462 Å	1.495 Å	1.307 Å	1.271 Å
	C_α	1.351 Å	1.371 Å	1.204 Å	1.223 Å
H3	All	2.165 Å	2.198 Å	1.761 Å	1.814 Å
	B/B	1.392 Å	1.475 Å	1.189 Å	1.279 Å
	C_α	0.997 Å	1.027 Å	0.755 Å	0.755 Å
Overall	All	2.110 Å	2.114 Å	1.957 Å	1.945 Å
	B/B	1.289 Å	1.309 Å	1.288 Å	1.287 Å
	C_α	1.234 Å	1.239 Å	1.245 Å	1.237 Å

factors (J. Pederson & A. R. Rees, unpublished work); (4) select order in which CDRs are to be constructed; and (5) model combining site according to Fig. 1.

As an indication of the success of this method, we show in Table 1 a comparison of r.m.s. deviations for the modelled and recent X-ray structure (P. D. Jeffrey, G. L. Taylor, R. Griest & A. R. Rees, unpublished work) of the anti-lysozyme antibody, GLOOP 2.

From these data and those of other antibodies recently completed and in progress, we are confident that, within the near future, the modelling of antibody-combining sites by a procedure in which the user is not required to make subjective structural judgments will become routine.

Metallo-Antibodies: a Dual Strategy for Design

(i) *Prediction of metal-binding sites*

We have used the methods summarized in the previous section to design a metal-binding site within the combining site of the anti-lysozyme antibody HyHEL-5 [18]. This antibody was chosen as a target for two reasons. First, the X-ray structure analysis of its complex with lysozyme has been carried out. Secondly, inspection of the structure suggested that regions of the antibody-combining site not engaged in antigen interactions might be suitable targets for protein engineering.

Metal-binding sites in proteins are often characterized by a hydrophilic region immediately around the metal position (owing to the presence of O, N or S atoms from the liganding groups) surrounded by a region of high hydrophobicity. This property of high 'hydrophobicity contrast' has suggested a way in which potential sites for metal binding could be identified in non-metallo-proteins. The use of atomic solvation parameters has previously been described by Eisenberg and co-workers [20], who later proposed [21] a hydrophobicity contrast function (c) of the following form:

$$c(r_i, \Delta\theta) = \sum (\Delta\theta_i \times r_i^2) - n\langle\Delta\theta\rangle\langle r^2\rangle$$

where r_i^2 is the square of the distance of the ith atom from the centre of a searching sphere, $\Delta\theta$ is a solvation parameter and n is the number of atoms in the sphere.

We have modified this function by applying an atomic hydrophobicity scale derived from atomic partial charge data rather than the atomic solvation parameters of Eisenberg & McLachlan [20]. A combination of this modification, with immersion of the target protein to be searched in a solvent box to eliminate spurious high contrast sites at the protein surface, has allowed development of an automated procedure that successfully identifies Ca^{2+}-, Cu^{2+}- and Fe^{2+}-binding sites in known metallo-proteins (D. S. Gregory, C. R. Martin & A. R. Rees, unpublished work). However, when this program is used to search for Zn^{2+}-binding sites, the success rate is somewhat lower. Zn^{2+}-binding sites tend to exhibit lower hydrophobic contrast than those of the 'harder' metals due to the preference for N-ligands (usually histidine). A new version of the program tackles this problem by first searching for appropriate cavities within the protein, since a metal-binding site must contain a cavity, the dimensions of which are defined by the liganding residues and the metal ion. All such cavities are then screened by the contrast function. The details of these algorithms and the results will be presented elsewhere.

(ii) *Knowledge-based design of metal-binding sites*

To introduce a metal-binding site into the combining site of an antibody, in such a way that antigen binding is retained, places a restriction on possible mutation sites. For example, CDR residues involved in antigen interactions cannot be mutated, except where the mutation is neutral as in, for example, the exchange of one hydrophobic residue for another in a region of 'loose' contact. Again, residues involved in CDR–CDR, CDR–framework or framework–framework interactions must be preserved. To test the likely effect of mutations, we have adopted a procedure whereby all substitutions are modelled by the protocol described earlier in this paper. In the particular case of a metal-binding site, this modelling is carried out in the presence and absence of the metal ion enabling the contribution of the metal to the stability of the mutant to be estimated.

Two approaches can be taken where the metal site is to be incorporated into CDRs: (1) the structural database (Brookhaven) can be searched for the presence of metal-binding loops with some similarity to one or more of the CDRs of the target antibody. A chimeric loop can then be constructed that contains the residual side-chains necessary to preserve antigen binding and CDR–CDR contacts and new residues that provide the metal-binding site; and (2) individual residues within a CDR can be mutated to known metal-binding groups (e.g. His, Cys, Glu, etc.) at positions within the CDR that enable formation of a metal-binding site with the appropriate geometry. In both instances, molecular modelling of the new CDRs is carried out to identify the energetically most favourable structure.

Applying the first approach, we examined the metal-binding environments of metallo-proteins the structures of which are known. Where at least three of a possible four liganding residues were contained within a single stretch of polypeptide chain, the loop structures were selected for computer grafting on to a suitably truncated CDR. An example of a metal-binding site from alcohol dehydrogenase that exhibits the desired characteristics is shown in Fig. 2. Next, a CDR was chosen that satisfied the criteria given in approach (1) above. The antibody chosen was the anti-lysozyme antibody HyHEL-5 [18], since one of its CDRs contained a region of six residues that neither contributed to lysozyme binding (from inspection of the structure) nor were involved in CDR–CDR interactions. In the next stage, the CDR was computer grafted in turn on to the different metal-binding loops and each chimera screened by the method of Martin *et al.* [17]. The hybrid selected on conformational criteria was finally modelled with and without the metal ion present. Fig. 3 shows the backbone conformations of a reconstructed light-chain CDR (L1) using parts of the HyHEL-5 structure and the Cd^{2+}-binding loop from metallothionein modelled according to Martin *et al.* [17], compared with the X-ray structure. From a single visual inspection it can be seen that the two CDRs have similar

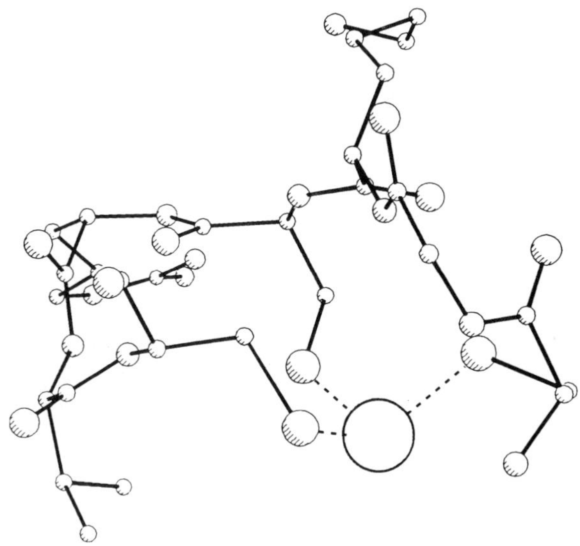

Fig. 2. *The conformation of residues 97–103 of liver alcohol dehydrogenase [25] showing the positions of the three cysteinyl ligands (97, 100, 103) and the zinc atom*

backbone conformations. Thus, it appears possible to replace whole segments of antibody CDRs, provided the regions concerned are not involved in interactions with antigen or play a structural role. This approach has been experimentally tested in our laboratory. First, to create wild-type HyHEL-5, for which the cDNA was not available, we CDR-grafted the HyHEL-5 sequences on to the framework of GLOOP-2, after suitable adjustment of CDR-framework residue pairs. Secondly, wild-type HyHEL-5–GLOOP-2 was expressed in frog oocytes [22] and showed identical binding affinity for

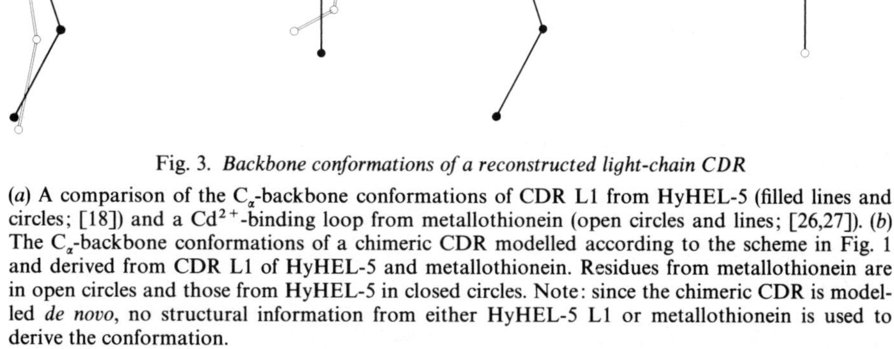

Fig. 3. *Backbone conformations of a reconstructed light-chain CDR*

(*a*) A comparison of the C_α-backbone conformations of CDR L1 from HyHEL-5 (filled lines and circles; [18]) and a Cd^{2+}-binding loop from metallothionein (open circles and lines; [26,27]). (*b*) The C_α-backbone conformations of a chimeric CDR modelled according to the scheme in Fig. 1 and derived from CDR L1 of HyHEL-5 and metallothionein. Residues from metallothionein are in open circles and those from HyHEL-5 in closed circles. Note: since the chimeric CDR is modelled *de novo*, no structural information from either HyHEL-5 L1 or metallothionein is used to derive the conformation.

lysozyme as authentic HyHEL-5 antibody (a gift from Dr S. Smith-Gill). Thirdly, a chimeric CDR-carrying metallo-sequence was constructed and expressed in the same way. Preliminary data have shown that the double chimeric antibody retains its ability to bind lysozyme with equal affinity to the wild-type chimera. Experiments on the metallo-properties of the antibody are in progress.

In conclusion, we believe that the methods under development in our laboratory and described in summary form in this paper will be important for a number of applications. It is likely that, in the very near future, it will be possible to model antibody-combining sites with high fidelity from a knowledge of the amino acid sequences of the variable regions only. We have also demonstrated that it is possible to replace segments of CDRs with totally different sequences encoding some other function and still retain antigen binding. This would not have been possible (except by random methods) without screening by molecular modelling. Finally, should it become routinely possible to engineer metal-binding sites into antibodies, this will undoubtedly find application in design of better biosensors and catalytic antibodies or perhaps for the preparation of metallo-antibodies for clinical use.

Note added in proof. Since this work was presented, a report of a metallo-antibody, constructed by mutagenesis of an anti-fluorescein antibody, has been published [*Science* (1990) **249**, 659–662].

We acknowledge support from the S.E.R.C. (J.C.), the M.R.C. (A.C.R.M.) and Amersham International (D.S. and D.G.).

References

1. Wu, T. T. & Kabat, E. A. (1970) *J. Exp. Med.* **132**, 211–250
2. Alzari, P. M., Lascombe, M.-B. & Poljak, R. J. (1987) *Annu. Rev. Immunol.* **6**, 555–580
3. Powell, M. J. & Hansen, D. E. (1989) *Protein Eng.* **3**, 69–75
4. Roberts, S., Cheetham, J. C. & Rees, A. R. (1987) *Nature (London)* **328**, 731–734
5. Riechmann, L., Clark, M., Waldmann, H. & Winter, G. (1988) *Nature (London)* **332**, 323–327
6. de la Paz, P., Sutton, B. J., Darsley, M. J. & Rees, A. R. (1986) *EMBO J.*, **5**, 415–425
7. Mainhart, C. R., Potter, M. & Feldmann, R. J. (1984) *Mol. Immunol.* **21**, 469–478
8. Smith-Gill, S. J., Mainhart, C. R., Lavoie, T. B., Feldmann, R. J., Droham, W. & Brooks, B. R. (1987) *J. Mol. Biol.* **194**, 713–724
9. Chothia, C., Lesk, A. M., Levitt, M., Amit, A. G., Mariuzza, R. A., Phillips, S. E. V. & Poljak, R. J. (1986) *Science* **233**, 755–758
10. Chothia, C. & Lesk, A. M. (1987) *J. Mol. Biol.* **196**, 901–917
11. Jones, T. A. & Thirup, S. (1986) *EMBO J.* **5**, 819–822
12. Brooks, B., Bruccoleri, R. E., Olafson, B. D., States, D. J., Swaminathan, S. & Karplus, M. (1983) *J. Comput. Chem.* **4**, 187–217
13. Moult, J. & James, M. N. G. (1986) *Proteins: Struct. Funct. Genet.* **1**, 146–163
14. Bruccoleri, R. E. & Karplus, M. (1987) *Biopolymers* **26**, 137–168
15. Fine, R. M., Wang, H., Shenkin, P. S., Yarmush, D. L. & Levinthal, C. (1986) *Proteins: Struct. Funct. Genet.* **1**, 342–362
16. Bernstein, F. C., Koetzle, T. F., Williams, G. J. B., Meyer, E. F., Brice, M. D., Rogers, J. R., Kennard, O., Shimanouchi, T. & Tasumi, M. (1977) *J. Mol. Biol.* **112**, 535–542
17. Martin, A. C. R., Cheetham, J. C. & Rees, A. R. (1989) *Proc. Natl. Acad. Sci. U.S.A.* **86**, 9268–9272
18. Sheriff, S., Silverton, E. W., Pedlan, E. A., Cohen, G. H., Smith-Gill, S. J., Finzel, B. C. & Davies, D. R. (1987) *Proc. Natl. Acad. Sci. U.S.A.* **84**, 8075–8079

19. Chothia, C., Lesk, A. M., Tramontano, A., Levitt, M., Smith-Gill, S. J., Air, G., Sheriff, S., Padlan, E. A., Davies, D. R., Tulip, W. R., Colman, P. M., Spinelli, S., Alzari, P. M. & Poljak, R. J. (1989) *Nature* (*London*) **342**, 877–883
20. Eisenberg, D. & McLachlan, A. D. (1986) *Nature* (*London*) **319**, 199–203
21. Eisenberg, D., Yamashita, M. & Eisenman, G. (1988) *ICSU Short Rep.: Adv. Gene Technol. Protein Eng. Product.* **8**, 3
22. Roberts, S. & Rees, A. R. (1986) *Protein Eng.* **1**, 59–66
23. Åqvist, J., van Gunsteren, M. F., Leifonmark, M. & Tapia, O. (1985) *J. Mol. Biol.* **183**, 461–477
24. Martin, A. C. R. (1990) D.Phil. Thesis, University of Oxford, U.K.
25. Plapp, B. V., Eklund, H., Jones, T. A. & Branden, E. I. (1983) *J. Biol. Chem.* **258**, 5537–5547
26. Furey, W. F., Robbins, A. H., Clancy, L. L., Winge, D. R., Wang, B. C. & Stout, C. D. (1986) *Science* **231**, 704–710
27. Schultze, P., Wörgötter, E., Braun, W., Wagner, G., Vasak, M., Kägi, J. H. R. & Wüthrich, K. (1988) *J. Mol. Biol.* **203**, 251–268

Subject Index

First and last page numbers of papers to which entries refer are given